BrightRED Study Guide

Curriculum for Excellence

N4

CHEMISTRY

Shona Scheuerl, Shona Wallace and Robert West

First published in 2015 by:
Bright Red Publishing Ltd
1 Torphichen Street
Edinburgh
EH3 8HX

A CIP record for this book is available from the British Library

ISBN 978-1-906736-44-7

With thanks to:
PDQ Digital Media Solutions Ltd, Bungay (layout) and Ailsa Morrison (copy-edit)
Cover design and series book design by Caleb Rutherford – e i d e t i c

Acknowledgements
Every effort has been made to seek all copyright-holders. If any have been overlooked, then Bright Red Publishing will be delighted to make the necessary arrangements.

Permission has been sought from all relevant copyright holders and Bright Red Publishing is grateful for the use of the following:

Robert West (p 2); Lisherwood/iStock.com (p 6); Chris Lofthouse (p 6); ValentynVolkov/iStock.com (p 8); Emilian Robert Vicol (CC BY 2.0)[1] (p 10); Steve Jurvetson (CC BY 2.0)[1] (p 10); Greenhorn1 (public domain) (p 11); Tavoromann (CC BY-SA 3.0)[2] (p 11); W. Oelen (CC BY-SA 3.0)[2] (p 11); lchemist-hp (CC BY 3.0)[3] (p 11); H. Michael Miley (CC BY-SA 2.0)[4] (p 11); jarsem/freeimages.com (p 14); Austin Kirk (CC BY 2.0)[1] (p 14); Caleb Rutherford e i d e t i c (p 19); NASA Expedition 6 crew (public domain) (p 21); Choba Poncho (public domain) (p 21); Dendrofil (public domain) (p 21); Roger McLassus (CC BY-SA 3.0)[2] (p 21); Izmaelt (CC BY-SA 3.0)[2] (p 21); Capt. John Yossarian (CC BY-SA 3.0)[2] (p 22); PetrS. (CC BY-SA 3.0)[2] (p 22); Stephanb (CC BY-SA 3.0)[2] (p 23); ginosphotos/iStock.com (p 23); elenabs/iStock.com (p 24); angelblue1/iStock.com (p 24); deyangeorgiev/iStock.com (p 24); Yannickcosta1 (CC BY-SA 3.0)[2] (p 26); Thomas Bjørkan (CC BY-SA 3.0)[2] (p 26); Chris Lofthouse (p 26); Robert West (p 27); kgreggain/freeimages.com (p 30); JustDavePhotography/iStock.com (p 30); Davide Guglielmo/freeimages.com (p 30); Gary Mirams (CC BY-SA 4.0)[5] (p 30); Arivumathi (CC BY-SA 3.0)[2] (p 30); Perytskyy/iStock.com (p 31); Andrew "FastLizard4" Adams (CC BY-SA 2.0)[4] (p 32); Three images by Caleb Rutherford e i d e t i c (p 33); Alcoholic Pyromaniacs (p 33); Timages/Dreamstime.com (p 34); Jacques-Louis David (public domain) (p 34); Ove Töpfer/freeimages.com (p 35); George Shuklin (public domain) (p 35); Health & Safety Symbols © Crown Copyright. Contains public sector information published by the Health and Safety Executive and licensed under the Open Government Licence' (p 35); Bomboman/Dreamstime.com (p 36); 4652 Paces (CC BY-ND 2.0)[6] (p 36); ARTPUPPY/iStock.com (p 37); Graeme Maclean (CC BY 2.0)[1] (p 38); Ranglen/Shutterstock.com (p 40); Jeppestown (CC BY-SA 2.0)[4] (p 40); Chris Lofthouse (p 40); Carlo Toffolo/Shutterstock.com (p 40); raining girl (CC BY 3.0)[3] (p 40); Mona Makela/Dreamstime.com (p 41); Serge Melki (CC BY 2.0)[1] (p 42); BlueBreezeWiki (CC BY-SA 3.0)[2] (p 42); Rufino Uribe (CC BY-SA 2.0)[4] (p 42); Myrabella (CC BY-SA 4.0)[5] (p 42); Michael Stocks/freeimages.com (p 43); Dutchscenery/Dreamstime.com (p 43); Annie Reynolds/PhotoLink (p 45); MeenaInc/freeimages.com (p 46); THOR (CC BY 2.0)[1] (p 48); hlphoto/iStock.com (p 48); Dan Taylr (CC BY 2.0)[1] (p 48); David Hurst (p 48); Shona Scheuerl (p 49); smartstock/iStock.com (p 50); Pakhnyushchyy/iStock.com (p 50); Cierpki/freeimages.com (p 50); www.Michie.ru (CC BY 2.0)[1] (p 50); Didriks (CC BY 2.0)[1] (p 50); freefoodphotos.com (p 50); karandaev/iStock.com (p 50); Richard Gailey (CC BY 2.0)[1] (p 50); Hajotthu (CC BY-SA 3.0)[2] (p 50); The Alcohol Education Trust (www.alcoholeducationtrust.org) (p 51); epSos .de (CC BY 2.0)[1] (p 52); Rachel Kramer (CC BY 2.0)[1] (p 52); Maja Dumat (CC BY 2.0)[1] (p 52); Valentyn75 (p 53); Mary Hutchison (CC BY 2.0)[1] (p 53); Jim Clark (CC BY 2.0)[1] (p 53); Vassil (public domain) (p 53); Chris Lofthouse (p 53); Paweł (CC BY-SA 2.0)[4] (p 53); Chris Lofthouse (p 56); webtreats (CC BY 2.0)[1] (p 56); jordachelr/iStock.com (p 56); Mike Warot (CC BY 2.0)[1] (p 56); seiki14/iStock.com (p 56); sedmak/iStock.com (p 56); Foodpictures/Shutterstock.com (p 56); Emilian Robert Vicol (CC BY 2.0)[1] (p 56); Gokhan Okur/Dreamstime.com (p 56); epSos .de (CC BY 2.0)[1] (p 56); Chris Lofthouse (p 57); James Cridland (CC BY 2.0)[1] (p 57); Serendipity Diamonds (CC BY-ND 2.0)[6] (p 57); Elena Elisseeva/Shutterstock.com (p 57); Caleb Rutherford e i d e t i c (p 58); Xavier13540 (public domain) (p 58); Leiem (CC BY-SA 4.0)[5] (p 58); Capt. John Yossarian (CC BY-SA 3.0)[2] (p 58); William Herron (CC BY-SA 2.0)[4] (p 59); Markus Grossalber (CC BY 2.0)[1] (p 59); Simon (CC BY-SA 3.0)[2] (p 60); saphon (CC BY-SA 1.0)[7] (p 60); Thegreenj (CC BY-SA 3.0)[2] (p 60); Anagoria (CC BY 3.0)[3] (p 60); James St. John (CC BY 2.0)[1] (p 60); Two photos by Chris Lofthouse (p 61); Arseniy P (CC BY-SA 3.0)[2] (p 61); Alchemist-hp (FAL 1.3)[9] (p 61); Schmid & Rauch (public domain) (p 62); Michela Simoncini (CC BY 2.0)[1] (p 62); lyng883 (CC BY 2.0)[1] (p 62); Renthal1969 (public domain) (p 63); Grafikę stworzył (CC BY-SA 3.0)[2] (p 63); Roger Alexander (CC BY 2.0)[1] (p 63); Huguette Roe/Shutterstock.com (p 64); Patrick McCall/Shutterstock.com (p 64); Dino Osmic/Shutterstock.com (p 64); cvilletomorrow (CC BY 2.0)[1] (p 66); Hispalois (CC BY-SA 3.0)[2] (p 66); Steve Anderson (CC BY 2.0)[1] (p 66); Denis Tabler/Shutterstock.com (p 66); Adam Kubalica (CC BY 2.0)[1] (p 66); Dana Rothstein/Dreamstime.com (p 66); rypson/iStock.com (p 67); Program Executive Office Soldier (public domain) (p 67); hroe/iStock.com (p 67); Sign taken from www.mysafetylabels.com (p 67); NASA/PhotoDisc (p 68); 夏天 (CC BY-ND 2.0)[6] (p 68); Dennis Jarvis (CC BY-SA 2.0)[4] (p 68); Isn't That Nice (CC BY 2.0)[1] (p 68); Chris Northwood (CC BY-SA 2.0)[4] (p 68); empire331/iStock.com (p 68); Davide Guglielmo/freeimages.com (p 69); NataliTerr/Shutterstock.com (p 70); nancybeetoo (CC BY 2.0)[1] (p 70); Ian Barbour (CC BY-SA 2.0)[4] (p 70); Bobby McKay (CC BY-ND 2.0)[6] (p 71); NASA/GSFC/SDO (CC BY 2.0)[1] (p 72); מודנר (CC BY-SA 3.0)[2] (p 72); Bork/Shutterstock.com (p 72); Stillwaterising (public domain) (p 72); Paul Nadar (public domain) (p 73); Wellcome Library, London (CC BY 4.0)[8] (p 73); Ilya Rabkin/Shutterstock.com (p 73); SchubPhoto/Shutterstock.com (p 73); Danny Nicholson (CC BY-ND 2.0)[6] (p 74); Søren Wedel Nielsen (CC BY-SA 3.0)[2] (p 74); Herge (public domain) (p 74); Søren Wedel Nielsen (CC BY-SA 3.0)[2] (p 74); Didaktische.Medien (CC BY-SA 3.0)[2] (p 74); Two photos by Søren Wedel Nielsen (CC BY-SA 3.0)[2] (p 74); PRHaney (CC BY-SA 3.0)[2] (p 75); Ildar Sagdejev (CC BY-SA 3.0)[2] (p 75); Steven Straiton (CC BY 2.0)[1] (p 76); tilo/iStock.com (p 76); Laura Riquelme/Shutterstock.com (p 76); Dave Crosby (CC BY-SA 2.0)[4] (p 76); ggw1962/Shutterstock.com (p 79); Markbob1968 (CC BY-SA 3.0)[2] (p 79); Cjp24 (CC BY-SA 3.0)[2] (p 79); Kasia75/iStock.com (p 80); Christian Ferrari/freeimages.com (p 85); Romangorielov/Dreamstime.com, Caleb Rutherford e i d e t i (p 87); Tomi/PhotoLink (p 89).

Cover image © Caleb Rutherford

Printed and bound in the UK by Charlesworth Press

CONTENTS

BRIGHT RED STUDY GUIDE: NATIONAL 4 CHEMISTRY

INTRODUCING NATIONAL 4 CHEMISTRY

This course allows you to develop a wide range of scientific and life skills that will equip you for a future of changing challenges. Its structure provides you with opportunities to develop and extend a wide range of chemistry-focused skills, while helping you to develop an understanding of chemistry's role in the scientific issues that affect society.

THE BENEFITS OF NATIONAL 4 CHEMISTRY

The course is split into four units, each with a 'real-life' theme. The course content will help you to develop knowledge and understanding of the chemistry around you. There will be plenty of opportunity to develop skills of scientific inquiry, as well as investigative and analytical thinking skills, within a chemistry context. Experimental work will allow development of planning and practical skills as well as building an awareness of safety considerations.

The National 4 course is a way to greatly enhance your understanding of the chemistry affecting your everyday life, while developing skills that will help you unravel the scientific issues affecting society and support you in many aspects of your life.

There is quite a lot to come to terms with in this course but, broken down as it is here, it is fairly straightforward. So, what is the structure?

INTERNAL ASSESSMENT: THE UNITS

There is considerable flexibility in the National 4 course in that there is no specified mandatory content. Knowing your abilities first hand, your teacher or lecturer will decide the most appropriate ways to cover the Key Areas of the course. The Key Areas that are covered during the course are assessed by your teachers or lecturers, who will be devising their own assessment programmes. They enjoy flexibility in how they conduct these assessments to give maximum variety and interest to the assessment process. Sometimes, they will be able to assess several outcomes at once; at other times, you will be given discrete tasks to perform, each assessing one skill.

How do these Key Areas operate?

They will be assessed by your teacher or lecturer according to SQA guidance.

The Key Areas are:

Chemical Changes and Structure
- Rates of reaction
- Atomic structure and bonding related to properties of materials
- Energy changes of chemical reactions
- Acids and bases

Nature's Chemistry
- Fuels
- Hydrocarbons
- Everyday consumer products
- Plants to products

Chemistry in Society

- Metals and alloys
- Materials
- Fertilisers
- Nuclear chemistry
- Chemical analysis

These Key Areas may be grouped together into three units: Chemical Changes and Structure, Nature's Chemistry and Chemistry in Society.

The Added Value Unit

The fourth unit is the Added Value Unit. In this unit you will draw on and apply the skills and knowledge you have learned during the course. You will carry out an in-depth investigation on an unfamiliar and/or integrated topic. The Added Value Unit is assessed through an assignment.

The assignment will be an in-depth study of a topical issue from a key area of the course chosen by you in agreement with your teacher or lecturer. The assignment will be assessed by your teacher or lecturer.

The assignment is carried out under controlled conditions. To prepare for the assessment you will choose, research/investigate an appropriate topic, focusing particularly on the applications and impact on society or the environment, and process the information.

During the assessment you will present evidence of:

- the issue being investigated and its relevance to the environment/society
- a selection of appropriate information from at least two relevant sources
- information displayed appropriately
- a description of the chemistry of the issue and its impact on the environment/society.

HOW WILL THIS GUIDE HELP YOU MEET THE CHALLENGES?

The aim of this book is to help you achieve success in the assessment of Key Areas by providing you with a suggested coverage of the Key Areas of the course. Helpful hints are provided throughout the book in the 'Don't Forget' features, while there are plenty of opportunities to practise applying your knowledge through 'Things to Do and Think About'. Some of the skills you will be expected to demonstrate are also covered in the book. These may be in with the relevant Key Areas or covered in separate sections.

CHEMICAL CHANGES AND STRUCTURE

REACTION RATES

The speed or rate of a chemical reaction is a measure of how fast the reaction occurs.

Different reactions have different rates. For example, the formation of crude oil from the remains of sea organisms is a very slow reaction while the burning and explosions which happen in fireworks are very fast reactions.

Crude oil – reactions took millions of years

VARIABLES AFFECTING THE RATE OF A REACTION

The reaction between magnesium metal and dilute hydrochloric acid is often used to study reaction rates. This reaction releases bubbles of hydrogen gas.

Concentration

In general, increasing the **concentration** of the reactants will lead to an increase in reaction rate.

Particle size

In general, increasing the **surface area** of a reactant will increase the rate of a reaction. Magnesium powder will react much faster with dilute hydrochloric acid than a piece of magnesium ribbon of the same mass.

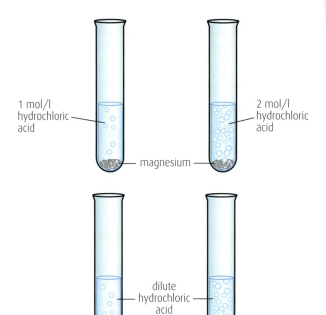

1 mol/l hydrochloric acid

2 mol/l hydrochloric acid

magnesium

dilute hydrochloric acid

magnesium ribbon

magnesium powder

Burning fireworks – reactions are over in seconds

Magnesium reacting with dilute hydrochloric acid

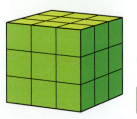

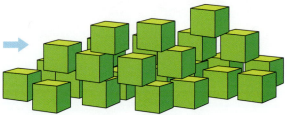

Consider a cube with a side length of 3 cm.

Each face has an area of 9 cm² (3 x 3) and since there are 6 faces the total surface area of the cube will be 54 cm².

Now split this cube into 27 little cubes. The total surface area is now 162 cm².

DON'T FORGET

The concentration of a solution is measured in moles per litre, mol/l. A 2 mol/l solution is twice as concentrated as a 1 mol/l solution.

DON'T FORGET

Crushing a lump of a solid will decrease the particle size of the lump but increases the surface area. Smaller particles have a larger surface area.

Temperature

In general, increasing the **temperature** of a reaction will increase the reaction rate.

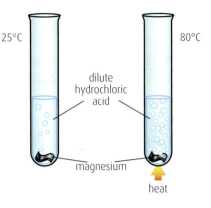

Catalysts

A **catalyst** is a substance which speeds up a reaction, but is chemically unchanged at the end of the reaction.

At the end of the reaction it is the same substance and it has the same mass as at the start of the reaction.

Consider the reaction between zinc metal and dilute sulfuric acid. The hydrogen gas produced can be seen as bubbles in the reaction mixture.

Copper metal acts as a catalyst for this reaction.

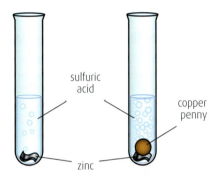

THINGS TO DO AND THINK ABOUT

1. Explain each of the following in terms of reaction rates.

   (a) Milk stays fresher for longer if it is kept in the fridge.

   (b) Large carrots take longer to cook than small carrots.

   (c) Charcoal on a barbecue glows more brightly when air is blown onto it.

2. Outline the steps necessary to prove that the copper penny is not used up in the reaction of zinc with sulfuric acid.

3. The table below shows the conditions used for six experiments in an investigation into the rate of reaction between magnesium and dilute hydrochloric acid.

Reaction	A	B	C	D	E	F
Concentration of hydrochloric acid (mol/l)	0.1	0.2	0.5	0.1	0.2	0.5
Temperature (°C)	20	20	20	30	30	30
Form of magnesium	ribbon	powder	powder	powder	ribbon	powder

   (a) In which experiment will the reaction be quickest?

   (b) Which two experiments should be compared to investigate the effect of temperature on the rate of the reaction?

   (c) In which experiment is the reaction slowest?

MONITORING REACTION RATES

During a chemical reaction the reactants change into products. This means that as the reaction proceeds the reactants are being used up and the products are being formed.

There are several methods of monitoring the rate of a reaction. All the methods rely on measuring how much of a reactant is used up or how much of a product is formed in a given period of time.

HOW TO MEASURE REACTION RATE

Reactions which produce a gas can be used to monitor the reaction rate.

One such reaction occurs between marble chips (a form of calcium carbonate) and dilute hydrochloric acid. The gas carbon dioxide is produced in this reaction.

Measuring the volume of gas produced

The reaction rate is monitored by collecting the carbon dioxide gas over water or in a gas syringe at suitable time intervals.

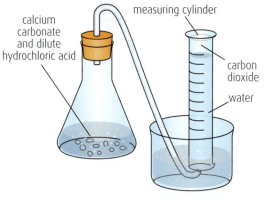
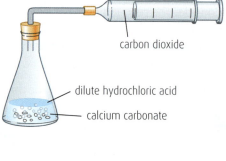

calcium carbonate and dilute hydrochloric acid

measuring cylinder

carbon dioxide

water

carbon dioxide

dilute hydrochloric acid

calcium carbonate

Measuring the mass of gas produced

If the reaction is set up on a balance as shown in the diagram, the mass of the apparatus will decrease as time passes due to carbon dioxide gas escaping.

Two methods of measuring the volume of gas produced in this reaction

REACTION RATE GRAPHS

Creating graphs from experiments helps us to visualise what happens to the reaction rate during the course of the reaction.

The results of an experiment to measure the rate of the reaction between marble chips and dilute hydrochloric acid are shown below.

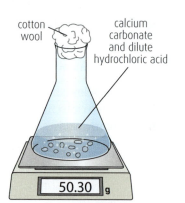

cotton wool

calcium carbonate and dilute hydrochloric acid

50.30 g

Time (s)	Volume of carbon dioxide gas produced (cm³)
0	0
20	40
40	56
60	62
80	64

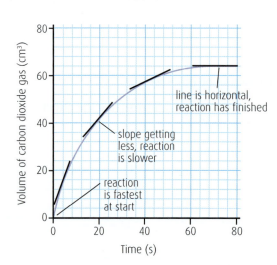

line is horizontal, reaction has finished

slope getting less, reaction is slower

reaction is fastest at start

Volume of carbon dioxide gas (cm³)

Time (s)

What can we learn from reaction rate graphs?

We can use graphs like these to find out information about the reaction. It is very important to realise two main points:

1. The rate of the reaction is related to the slope (gradient) of the line. The steeper the gradient of the line, the faster the reaction is at that time.

2. The reaction has stopped when the line is horizontal – it has a zero gradient.

Some questions and answers related to this graph are shown in the table.

Question	Acceptable answer
When was the reaction fastest?	At the beginning
At what time did the reaction stop?	Any time from 69 s to 71 s
What volume of gas was produced after 10 seconds?	Any volume from 27 cm³ to 29 cm³
What was the total volume of gas produced in the reaction?	64 cm³

Changing the reaction conditions

Two lines are plotted on the graph on the right. Each line represents a different set of conditions for the reaction between calcium carbonate and dilute hydrochloric acid.

Experiment B has the steeper line which indicates that the reaction conditions produced a faster reaction rate.

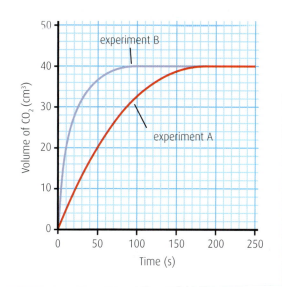

We can confirm this by looking at when each reaction stops producing carbon dioxide gas. Experiment A reaches this point at around 175 s while experiment B is complete at around 80 s.

The increased rate of experiment B could be due to:

- the use of a catalyst
- a higher reaction temperature
- an increased concentration of hydrochloric acid
- using the same mass of calcium carbonate with a smaller particle size (larger surface area).

THINGS TO DO AND THINK ABOUT

1. The graph shows the volume of oxygen collected during a chemical reaction carried out at 45°C.

   (a) What volume of oxygen was produced in the first 10 s?

   (b) How many seconds did it take for the reaction to stop?

   (c) What was the total volume of oxygen produced in this reaction?

   (d) Copy the graph and draw a dotted line to represent the same reaction carried out at 75°C.

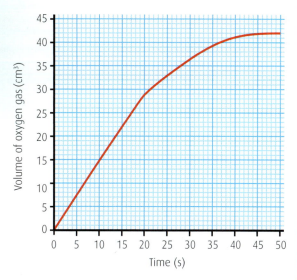

ATOMIC STRUCTURE – THE PERIODIC TABLE

ELEMENTS

An **element** is a substance which contains only one type of atom.

As of 2014, there are 118 known elements each with a different type of atom. Just as the 26 letters of the alphabet combine to make all the words we use, the chemical elements combine to make all the known chemical compounds. Elements can be thought of as the building blocks of all substances.

All the atoms in these copper pipes are the same type of atom – copper atoms.

THE MODERN PERIODIC TABLE

All the known elements are found in the Periodic Table. Each element has a different symbol and **atomic number**. The elements are placed into the Periodic Table in order of increasing atomic number. The Periodic Table is arranged in horizontal rows called **periods** and vertical columns called **groups**.

Features of the Periodic Table

- The majority of the elements are metals.
- Most elements are solids at room temperature; a few are gases and two, mercury and bromine, are liquids.
- The groups are numbered from 1 to 7. The final group is labelled with a zero.
- The transition metals lie between groups 2 and 3.

All the atoms in this diamond (a form of carbon) are the same type of atom – carbon atoms.

DON'T FORGET

Elements are made from one TYPE of atom, not from just one atom.

DON'T FORGET

A compound is a substance which contains at least TWO different types of atom joined together.

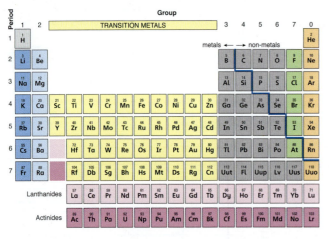

TASK

Select an element with an atomic number between 1 and 36 and find out all you can about it. You might include some or all of the following:

- how the element and/or its compounds have affected society or the environment
- symbol and atomic number
- who discovered the element
- date of discovery
- the main sources of the element
- uses.

Use the information you have found to create a poster or a PowerPoint presentation and present your element to the rest of your class.

DON'T FORGET

If the element symbol consists of one letter it is written as a capital. If the symbol has two letters, the first letter is a capital and the second letter is not.

Another important feature of the Periodic Table is that elements with similar chemical properties are grouped together.

Group 1 – The alkali metals

Li	Lithium
Na	Sodium
K	Potassium
Rb	Rubidium
Cs	Caesium
Fr	Francium

As with all the families of elements, the alkali metals are not identical, but they do show similar chemical properties.

The alkali metals are all soft metals which are shiny when freshly cut, but lose their shininess when exposed to air as a layer of oxide forms on the surface:

metal + oxygen → metal oxide

For this reason the metals are stored in liquid paraffin oil to help prevent contact with the air or with water.

The elements in group 1, with the exception of hydrogen, are called the 'alkali metals' because they react with water to form alkaline solutions.

For example, sodium reacts vigorously with water to produce the alkali sodium hydroxide. Hydrogen gas is also made in this reaction.

sodium + water → sodium hydroxide + hydrogen (an alkali)

Sodium metal stored in oil

Sodium reacting with water

Group 7 – The halogens

F
Cl
Br
I
At

The halogens are reactive non-metallic elements. They are all toxic and undergo similar chemical reactions.

Fluorine and chlorine are gases, bromine is a liquid and iodine is a solid. Astatine is a radioactive element which does not occur naturally on Earth.

The halogens and their compounds have many uses. Fluorine compounds are used in toothpaste to help prevent cavities. The most common use of chlorine is in drinking water as it can kill harmful bacteria. Bromine is used in dyes and medicines and iodine is used as an antiseptic.

Chlorine, bromine, iodine

Group 0 – The Noble gases

The group 0 elements are all colourless gases. They are extremely unreactive and form almost no known chemical compounds. They are used in lasers, lighting and airships.

He   Ne   Ar   Kr   Xe

The helium in this airship is much less dense than air.

THINGS TO DO AND THINK ABOUT

1. Are the following statements true or false?

- Calcium is an alkali metal.
- Hydrogen is an alkali metal.
- There are seven noble gases.
- Fluorine is the least reactive halogen.
- The noble gases form compounds easily.
- The alkali metals are very unreactive.
- The noble gas argon is used in light bulbs.
- The alkali metals are found in group seven.
- Sulfur is a member of the halogen family.
- Iodine is the only element which is a liquid at room temperature.

2. The elements 114 (Fl) and 116 (Lv) were given their names by the International Union of Pure and Applied Chemistry in May 2012. Try to find out how the names of these elements were decided.

ATOMIC STRUCTURE – THE STRUCTURE OF THE ATOM

SUB-ATOMIC PARTICLES

All solids, liquids and gases are made of atoms. Chemists now know a lot of information about the structure of atoms. The current atomic model indicates that an atom is made up of three even smaller **sub-atomic particles** called **protons**, **neutrons** and **electrons**.

Every atom has an extremely small **nucleus** containing protons and neutrons. The electrons are found outside the nucleus in regions of space called electron shells or energy levels.

Target diagrams are simplified drawings of atoms. These are often used to show the structure of an atom.

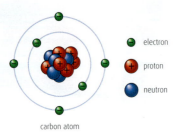

carbon atom

Target diagram of a carbon atom

All three sub-atomic particles are different from each other.

Protons – positively charged particles with a mass of 1 atomic mass unit.

Neutrons – neutral particles (they have no electric charge) with a mass of 1 atomic mass unit.

Electrons – negatively charged particles with almost zero mass.

Particle	Position	Relative mass	Charge
Proton	Nucleus	1	+1
Neutron	Nucleus	1	0
Electron	Outside the nucleus	Almost zero	−1

Relative mass

It is clear from the table that protons and neutrons are much heavier than electrons. To balance the mass of one proton or one neutron would require around 1830 electrons.

This means that the mass of an atom is concentrated in the nucleus. When the mass of an atom is calculated, the mass of any electrons present is ignored.

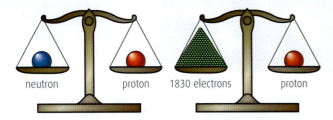

neutron   proton   1830 electrons   proton

Charge

Protons are positively charged particles, electrons are negatively charged particles and neutrons are particles with no electric charge.

Atoms are electrically neutral because the number of positively charged protons equals the number of negatively charged electrons.

nucleus

proton   neutron   electron

Consider the lithium atom shown below.

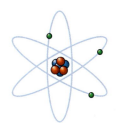

 Charge on 3 protons = 3 × (+1) = +3

 Charge on 3 electrons= 3 × (-1) = -3
Total charge = (+3) + (-3) = 0

ATOMIC NUMBER AND MASS NUMBER

Each element in the Periodic Table has its own **atomic number**. The atomic number of an element is equal to the number of protons in its nucleus. The Periodic Table arranges the elements in order of increasing atomic number. The first element, with an atomic number of one, is hydrogen. All hydrogen atoms have one proton in their nucleus.

Every atom also has a **mass number,** which is defined as the total number of protons and neutrons in its nucleus.

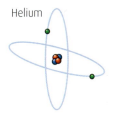
Helium

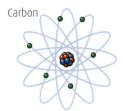
Carbon

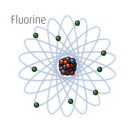
Fluorine

Helium   Atomic number is 2 as the atom has two protons. The mass number of this atom is 4 (2 protons and 2 neutrons).

Carbon   Atomic number is 6 as the atom has 6 protons. The mass number of this atom is 12 (6 protons and 6 neutrons).

Fluorine   Atomic number is 9 as the atom has 9 protons. The mass number of this atom is 19 (9 protons and 10 neutrons).

 THINGS TO DO AND THINK ABOUT

1. An atom of an element is represented by the diagram below.

    (a) What is the atomic number of this atom?

    (b) What is the mass number of this atom?

    (c) Is this atom a metal or a non-metal?

    (d) Explain why this atom is electrically neutral.

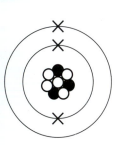

● = protons
○ = neutrons
✗ = electrons

2. An atom has 15 protons and 16 neutrons.

    (a) What is the atomic number of this atom?

    (b) What is the mass number of this atom?

    (c) Which element has atoms with 15 protons?

3. An atom of sodium has a mass number of 23.

    How many protons and neutrons does this sodium atom have?

NAMES AND FORMULAS 1

NAMING COMPOUNDS

All the substances in the universe are made from around 118 chemical elements. When two or more elements react together their atoms combine and new chemical compounds are formed.

The **common name** of a compound seldom reveals the elements it contains. However, the **chemical name** is based on the elements the compound is made from.

For example, table salt has the chemical name sodium chloride. It is easy to see from the chemical name that the compound contains the elements sodium and chlorine.

Salt – sodium chloride

Name ending...

...IDE	...ATE/ITE
Usually indicates the compound contains TWO ELEMENTS	Usually indicates the compound contains THREE ELEMENTS, one of which is ALWAYS OXYGEN

Examples	Examples
Copper bromide – copper and bromine Calcium oxide – calcium and oxygen Lithium chloride – lithium and chlorine	Magnesium nitrate – magnesium, nitrogen and oxygen Sodium sulfite – sodium, sulfur and oxygen Potassium chlorate – potassium, chlorine and oxygen

Talcum powder – magnesium silicate

TASK

Create a table to show the name and symbol of the elements present in each of the following compounds: sodium bromide, copper sulfate, iron chloride, potassium hydroxide, lithium carbonate, magnesium oxide, hydrogen chloride, zinc sulfide, chromium nitrate, nickel iodide.

WRITING FORMULAS FROM STRUCTURES AND NAMES

The chemical formula of a compound indicates the ratio of the particles which make up the compound.

EXAMPLE

Water is a chemical compound formed from the elements hydrogen and oxygen. The chemical name for water is hydrogen oxide. The chemical formula of water is H_2O. This formula indicates that each water molecule is made from two hydrogen atoms and one oxygen atom.

DON'T FORGET

The word hydroxide indicates hydrogen and oxygen. This means that hydroxide compounds will contain at least three elements even though their names end with –ide. Sodium hydroxide contains sodium, hydrogen and oxygen. Watch out for this exception to the rule.

DON'T FORGET

"ite"/"ate" indicates oxygen is present in a compound.

DON'T FORGET

When writing chemical formulae the number one is assumed and usually not written as part of the formula. For example: the formula of water is written as H_2O and not H_2O_1.

Formulas from structures

Chemists often draw the structure of molecules to show the atoms they contain. Sometimes lines are drawn between the atoms to show the chemical bonds which hold the atoms together. Consider the examples in the table.

Formulas from names

Sometimes the name of a compound may contain a prefix like mono, di or tri. The table shows the meaning of several commonly used prefixes. When this happens the formula of the compound can be found directly from its name.

Prefix	mono	di	tri	tetra	penta	hexa
Meaning	1	2	3	4	5	6

EXAMPLE

The formula of dinitrogen monoxide can be written straight from its name. The name indicates that the compound contains two nitrogen atoms and one oxygen atom and so its formula is N_2O.

More examples include:

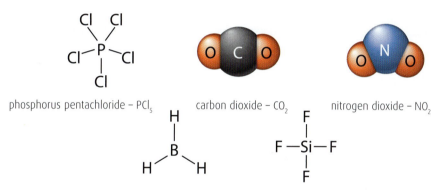

phosphorus pentachloride – PCl_5 carbon dioxide – CO_2 nitrogen dioxide – NO_2

boron trihydride – BH_3 silicon tetrafluoride – SiF_4

Structure	Formula
	H_2O
	$SiCl_4$
	SO_2
	C_2H_6O
	C_7H_{16}
	CCl_2F_2

THINGS TO DO AND THINK ABOUT

1. Write the formula of the following compounds.
   (a) Dinitrogen pentoxide   (b) Sulfur hexafluoride   (c) Phosphorus trichloride
   (d) Carbon disulfide   (e) Carbon monoxide.

2. Write the formula for the following compounds:
   (a)   (b)   (c) H — Br   (d)   (e)

   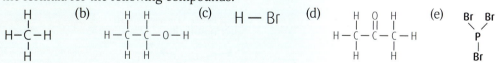

3. Caffeine is a stimulant found in tea, coffee and some fizzy drinks. The molecular formula of caffeine is $C_8H_{10}N_4O_2$.
   (a) How many different elements are in caffeine?
   (b) How many atoms make up one caffeine molecule?

4. Hydrogen cyanide is a very toxic gas. A possible structure of a hydrogen cyanide molecule is shown.
   (a) What is the formula of hydrogen cyanide?
   (b) Explain why chemistry students might expect hydrogen cyanide to contain only two elements.

   $$H — C \equiv N$$

NAMES AND FORMULAS 2

WHAT ARE VALENCY NUMBERS?

In a chemical compound the atoms of the elements are held together by chemical bonds. As mentioned earlier, diagrams of compounds often show these bonds as solid lines between the atoms.

$$
\begin{array}{l}
\text{Br} \quad \text{Br} \\
\quad\;\; \text{P} \\
\quad\;\; | \\
\quad\;\; \text{Br}
\end{array}
\qquad
\text{Cl}-\overset{\displaystyle \text{Cl}}{\underset{\displaystyle \text{Cl}}{\text{C}}}-\text{Cl}
\qquad
\text{Cl}-\overset{\displaystyle \text{F}}{\underset{\displaystyle \text{F}}{\text{C}}}-\text{Cl}
\qquad
\text{H}-\text{Br}
$$

$$
\text{H}-\overset{\displaystyle \text{H}}{\underset{\displaystyle \text{H}}{\text{C}}}-\text{H}
\qquad
\text{N}\equiv\text{N}
\qquad
\text{H}-\overset{\displaystyle \text{H}}{\underset{\displaystyle \text{H}}{\text{C}}}-\text{O}-\text{H}
$$

The atoms of most elements usually form the same number of bonds when they make chemical compounds. Looking at the structures shown we can see that hydrogen, chlorine, fluorine and bromine form one bond, oxygen forms two bonds, phosphorus and nitrogen form three bonds and carbon forms four bonds.

The **valency** or **valence number** of an atom can be defined as the number of bonds the atom can make. Valency is often called the combining power of an atom.

Valency and the Periodic Table

The valency number of an element can be found from the groups in the Periodic Table.

Elements with variable valencies

Some elements, particularly the transition metals, can have more than one valency number. The names of transition metal compounds usually indicate the valency of the metal with a **Roman numeral** after the element's name.

Copper(I) sulfide – copper has a valency of one in this compound.
Copper(II) oxide – copper has a valency of two in this compound.
Iron(III) hydroxide – iron has a valency of three in this compound.

USING VALENCY NUMBERS

Valency rules are a step-by-step process that enables the formulae of many compounds to be written. The first two examples show how to write the correct formulas for lithium oxide and aluminium chloride.

Group number	1	2	3	4	5	6	7	0
Valency number	1	2	3	4	3	2	1	0

EXAMPLES 1 & 2

	Lithium oxide		Aluminium chloride
Select symbols	Li ⟶ O		Al ⟶ Cl
Cross valency numbers	1 ✕ 2		3 ✕ 1
Write the formula	Li_2O		$AlCl_3$

DON'T FORGET

The number 1 is usually omitted in a chemical formula.

The two examples which follow demonstrate the idea of "cancelling down" valency numbers:

EXAMPLES 3 & 4

	Magnesium oxide		Silicon sulfide
Select symbols	Mg ⟶ O		Si ⟶ S
Cross valency numbers	2 ✕ 2		4 ✕ 2
Cancel down valencies	Mg_1 O_1		Si_1 S_2
Write the formula	MgO		SiS_2

DON'T FORGET

Valency numbers are always cancelled down to give the simplest whole numbers in the formula. For example, valencies of 4 and 2 cancel to 2 and 1, and valencies of 3 and 3 cancel to 1 and 1.

The two examples which follow consider the formulas of compounds when the name of the compound contains a Roman numeral:

EXAMPLES 5 & 6

	Copper(I) oxide		Nickel(II) chloride
Select symbols	Cu ⟶ O		Ni ⟶ Cl
Cross valency numbers	1 ✕ 2		2 ✕ 1
Write the formula	Cu_2O		$NiCl_2$

Look again at example 6. A common incorrect answer for the formula of nickel(II) chloride is Ni_2Cl. You can see that this formula arises because the student has failed to cross over the valency numbers and simply written the formula in the same order as suggested by the name of the compound.

Two final examples

Study these carefully to make sure you understand how to apply the basic valency rules:

EXAMPLES 7 & 8

	Iron(III) sulfide	Aluminium nitride
Select symbols	Fe S	Al N
Cross valency numbers	3 ⤬ 2	3 ⤬ 3
Cancel valencies	(not possible)	Al_1 N_1
Write the formula	Fe_2S_3	AlN

⊚ TASK   Valency Pictures

Writing chemical formulae can be a difficult skill to master. Some teachers use "valency pictures" to help their students understand this topic.

A valency picture shows the symbol of the element and its valency number, for example:

Group 1 Lithium: valency = one [Li]   Group 6 Sulfur: valency = two [S]

When the elements are joined together to form a compound the picture should have no spare bonds and elements must not join to themselves.

The valency picture for lithium sulfide is

This shows the formula of lithium sulfide is Li_2S

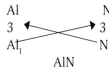

Similarly the valency picture of calcium chloride is

This shows the formula of calcium chloride is $CaCl_2$

Draw your own valency pictures for the elements hydrogen, sodium, magnesium, aluminium, carbon, chlorine and oxygen. Use your pictures to create a poster or PowerPoint which shows how they can be used to find the formulae of the compounds carbon dioxide, hydrogen oxide, aluminium chloride, magnesium oxide and sodium chloride.

THINGS TO DO AND THINK ABOUT

1. What is the valency number of each of the following elements?

   (a) Potassium
   (b) Silicon
   (c) Iodine
   (d) Magnesium
   (e) Aluminium
   (f) Oxygen

2. Use the valency rules to write the formulas of the following compounds.

   (a) Lithium bromide
   (b) Magnesium nitride
   (c) Silicon oxide
   (d) Strontium chloride
   (e) Aluminium oxide
   (f) Sodium sulfide
   (g) Mercury(II) oxide
   (h) Chromium(III) chloride
   (i) Cobalt(II) chloride
   (j) Manganese(V) oxide

3. Some chemical formulae are known not to conform to the normal valency rules. Write the formulas of the compounds below using only their names and then decide if they break the valency rules.

   (a) Phosphorus trichloride
   (b) Carbon monoxide
   (c) Sulfur dioxide
   (d) Dinitrogen oxide

FORMULA MASS

As you learned on pages 12 to 13 all atoms have mass. On the atomic mass scale hydrogen atoms have a relative atomic mass of 1. Oxygen atoms have a relative atomic mass of 16. This means that one oxygen atom has 16 times the mass of a hydrogen atom.

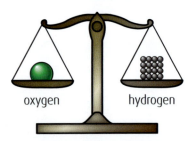

oxygen   hydrogen

One oxygen atom and 16 hydrogen atoms on the balance pans

The **formula mass** of a compound can be found from its formula if we know the relative atomic mass (RAM) of all the atoms in the compound.

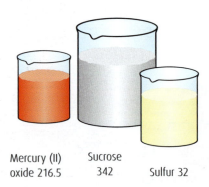

Mercury (II)   Sucrose
oxide 216.5   342   Sulfur 32

Formula mass of different substances

Element	Symbol	Relative atomic mass
aluminium	Al	27
argon	Ar	40
bromine	Br	80
calcium	Ca	40
carbon	C	12
chlorine	Cl	35·5
copper	Cu	63·5
fluorine	F	19
gold	Au	197
helium	He	4
hydrogen	H	1
iodine	I	127
iron	Fe	56
lead	Pb	207
lithium	Li	7
magnesium	Mg	24·5
mercury	Hg	200·5
neon	Ne	20
nickel	Ni	58·5
nitrogen	N	14
oxygen	O	16
phosphorus	P	31
platinum	Pt	195
potassium	K	39
silicon	Si	28
silver	Ag	108
sodium	Na	23
sulfur	Si	32
tin	Sn	118·5
zinc	Zn	65·5

Calculate the formula mass of calcium bromide, $CaBr_2$

Step 1 – Identify the number of atoms $(1 \times Ca) + (2 \times Br)$

mass of calcium

two bromines in formula   mass of bromine

one calcium in formula

Step 2 – Change the symbols to the relative atomic mass $(1 \times 40) + (2 \times 80)$

Step 3 – Do the arithmetic   40   + 160

Step 4 – Final answer   200

DON'T FORGET

The formula mass is found by adding together all the relative atomic masses of the atoms or ions present in the substance.

DON'T FORGET

Give each element its own set of brackets at Step 1 in formula mass calculations.

EXAMPLE 2

Calculate the formula mass of sodium oxide, Na_2O

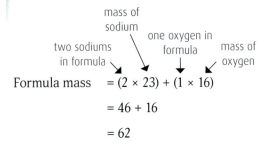

Formula mass $= (2 \times 23) + (1 \times 16)$

$= 46 + 16$

$= 62$

EXAMPLE 4

Calculate the formula mass of glucose, $C_6H_{12}O_6$

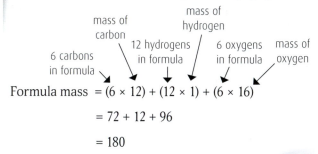

Formula mass $= (6 \times 12) + (12 \times 1) + (6 \times 16)$

$= 72 + 12 + 96$

$= 180$

EXAMPLE 3

Calculate the formula mass of iron(II) sulfide, FeS

Formula mass $= (1 \times 56) + (1 \times 32)$

$= 56 + 32$

$= 88$

THINGS TO DO AND THINK ABOUT

1. Calculate the formula mass of the following compounds

   (a) NaCl   (b) H_2O (c) CO_2 (d) $MgCl_2$ (e) Al_2O_3

   (f) MgO   (g) $SiBr_4$ (h) C_2H_6O (i) K_2S (j) PBr_5

2. Which of the following is the formula mass of SO_2?

   (a) 32   (b) 48   (c) 64   (d) 80

3. What is the formula mass of the compound on the right?

   (a) 47.5   (b) 190   (c) 83.5   (d) 154

4. Which of the following has the largest formula mass?

   (a) Br_2 (b) O_2 (c) H_2 (d) N_2

5. Which of the following has the smallest formula mass?

   (a) NaBr   (b) NaF   (c) NaCl   (d) NaI

6. Aiden calculated the formula mass of four different compounds. Unfortunately he mixed up the compounds and their formula masses when he entered his results in the table shown below.

Compound	Formula mass
SO_2	34
CaO	87
H_2S	64
LiBr	56

Copy and complete the table to show the correct formula mass for each compound.

IONIC AND COVALENT BONDING

COMPOUND NAMES

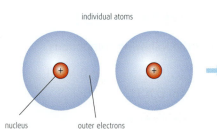

All chemical compounds are classified as either ionic or covalent. Ionic compounds are named as such because they are composed of ions and the ions are held together by ionic bonds. Many covalent compounds are made of molecules and the atoms in the compound are held together by covalent bonds.

Ionic compounds are usually produced when a metal element combines with a non-metal element. Covalent compounds contain only non-metal elements.

Knowing the name of a compound gives a good indication of the type of bonding present.

Name of compound	Type of bonding
sodium chloride	ionic
lithium carbonate	ionic
nitrogen hydride	covalent
phosphorus bromide	covalent

BONDING AND ELECTRONS

Chemical compounds are formed when elements combine and their atoms or ions bond together. We know from studying the structure of atoms that an atom has a central nucleus surrounded by electrons. Chemists believe that when atoms collide and react it is the electrons that are involved.

In the formation of an ionic compound, electrons are transferred from the metal atom to the non-metal atom. This results in the formation of charged particles called ions. In general, metal atoms form positive ions and non-metal atoms form negative ions.

In the formation of covalent compounds electrons are shared between the atoms and this results in the formation of molecules. Molecules tend to be electrically neutral.

DON'T FORGET

The name of a compound only gives an indication of the type of bonding present. To know for sure the properties of the compound must be known.

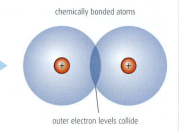

individual atoms

chemically bonded atoms

nucleus   outer electrons

outer electron levels collide

BONDING AND PROPERTIES

Ionic and covalent compounds have different properties due to their chemical bonding and structure. A compound is ionic if it has properties which are consistent with ionic bonding. The properties of a compound are the best way to classify it.

Melting point and boiling point

At its melting point a substance will change from a solid to a liquid and at its boiling point a substance will change from a liquid to a gas.

Ionic compounds have high melting and boiling points and as a result they tend to be crystalline solids at room temperature. Covalent substances, which are made of molecules, have low melting and boiling points and as a result they can exist as solids, liquids or gases at room temperature.

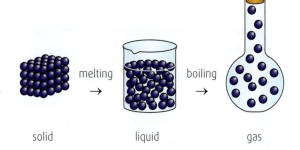

solid   melting →   liquid   boiling →   gas

Electrical conductivity

An electric current is a flow of charged particles. Substances which allow a current to flow are called conductors and substances which do not allow a current to flow are called non-conductors or electrical insulators.

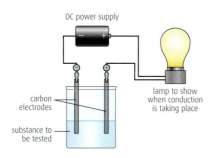

DC power supply

carbon electrodes

substance to be tested

lamp to show when conduction is taking place

One way of testing the electrical conductivity of a compound is shown. If the compound is a conductor, then a current will flow around the circuit and the bulb will light.

Results from experiments on conductivity reveal that covalent compounds do not conduct in any state. Ionic compounds do not conduct when in the solid state, but they do conduct if they are molten (melted) or when they are in aqueous solution (dissolved in water).

Substance	Bonding	Melting point (°C)	Boiling point (°C)	State at 25°C	Appearance
Sodium chloride	ionic	801	1413	solid	sodium chloride crystals
Copper(II) chloride	ionic	498	995	solid	copper(II) chloride crystals
Paraffin wax	covalent	47	380	solid	solid wax
Water	covalent	0	100	liquid	water
Nitrogen dioxide	covalent	-11	21	gas	brown nitrogen dioxide gas

Deciding type of bonding

If no other information is available it is reasonable to deduce the bonding in a compound from the elements it contains. However, it is much more reliable to use the properties of the compound to determine whether it is ionic or covalent.

Type of substance \ State	Solid	Molten	Aqueous solution
Covalent	insulator	insulator	insulator
Ionic	insulator	conductor	conductor

THINGS TO DO AND THINK ABOUT

1. Which of the following compounds are ionic and which are covalent?
   Magnesium chloride, lithium sulfate, nitrogen hydride, carbon chloride, potassium bromide, sodium nitrate, calcium oxide, hydrogen oxide.

2. The table gives information about the electrical conductivity and the melting points of several compounds.

Substance	Conducts as a solid	Conducts as a liquid	Melting point (°C)
A	no	yes	963
B	no	no	47
C	no	no	180
D	no	yes	765

   (a) Which two substances are ionic?
   (b) Which two substances are made of molecules?
   (c) Which substances are covalent?
   (d) Which substance would need the most energy to be turned from a solid into a liquid?

3. The grid shows the names and formulas of some elements and compounds.

copper Cu	methane CH_4	sulfur S
sodium chloride NaCl	carbon dioxide CO_2	calcium fluoride CaF_2

   (a) Name the ionic compounds.
   (b) Name the compounds made of molecules.
   (c) Name the elements.
   (d) Which two compounds are likely to have high melting points?
   (e) Which two compounds do not conduct electricity in any state?

CHEMICAL REACTIONS

ENERGY CHANGES

In a chemical reaction substances react together and are changed into new substances. In addition, most chemical reactions also involve an energy change.

Exothermic reactions

Any reaction which releases energy to the surroundings is called an **exothermic** reaction and this is usually accompanied by a rise in temperature.

EXAMPLE 1

Magnesium metal burns vigorously in oxygen releasing both heat and light energy.

EXAMPLE 2

The thermite reaction is a highly exothermic reaction between aluminium and iron(III) oxide. Temperatures well over 2000°C are reached during this reaction which produces molten iron as one of the products. The thermite reaction is often used to weld railways lines together.

Endothermic reactions

Endothermic reactions can be thought of as the opposite of exothermic reactions because they take in energy from the surroundings. When this happens the reaction mixture and the surroundings become colder.

An endothermic reaction is responsible for the cooling effect of certain types of sports injury cool packs. They consist of a large bag containing one reactant and an inner bag containing the second reactant. The inner bag is broken to allow the chemicals to react and an immediate drop in temperature is produced.

CHEMICAL EQUATIONS

When a chemical reaction occurs it can be described by an equation. The equation shows the **reactants** (the starting substances) and the **products** (the substances that are made). The reactants and products are separated by an arrow with the reactants on the left-hand side and the products on the right-hand side.

The chemicals involved in the reaction can be represented by words or by their chemical symbols or formulas. Consider the two exothermic reactions above.

EXAMPLE 1 MAGNESIUM BURNING

When magnesium burns it reacts with oxygen gas to produce the compound magnesium oxide.

The word equation for this reaction is:

magnesium + oxygen → magnesium oxide

A **formula equation** shows the correct formulas of all the substances involved in the reaction.

The formula equation for this reaction is:

$$Mg + O_2 \rightarrow MgO$$

You will notice that the formula of oxygen is written as O_2 and not simply as O. Oxygen atoms form diatomic molecules. In a diatomic molecule the atoms are joined together in pairs – they are two-atom molecules. There are six other elements which have diatomic molecules. The chemical formula of a diatomic element is therefore written as X_2 where X is the symbol of the element.

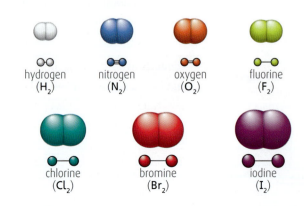

hydrogen (H_2)   nitrogen (N_2)   oxygen (O_2)   fluorine (F_2)

chlorine (Cl_2)   bromine (Br_2)   iodine (I_2)

DON'T FORGET

Chemical equations use arrows (→) not equals signs (=) to separate the reactants and products.

EXAMPLE 2 THE THERMITE REACTION

When aluminium powder and solid iron(III) oxide react there is a huge release of energy. Molten iron and solid aluminium oxide are produced.

The word equation for this reaction is:
aluminium + iron(III) oxide → iron + aluminium oxide

The formula equation for this reaction is: $Al + Fe_2O_3 \rightarrow Fe + Al_2O_3$

copper(II) sulfate,
$CuSO_4(s)$

State symbols

Chemicals can exist as solids, liquids or gases. Reactions often take place in aqueous solution (the substance is dissolved in water). State symbols can be added to the right of a chemical formula to indicate the state of the substance.

State symbols: (s) = solid   (l) = liquid   (g) = gas   (aq) = aqueous solution

copper(II) sulfate,
$CuSO_4(aq)$

For example, hydrogen gas would be written as $H_2(g)$, solid magnesium would be written as Mg(s) and liquid water would be written as $H_2O(l)$.
State symbols can be included in equations.

EXAMPLE 1

Burning magnesium
$Mg(s) + O_2(g) \rightarrow MgO(s)$

EXAMPLE 2

The thermite reaction
$Al(s) + Fe_2O_3(s) \rightarrow Fe(l) + Al_2O_3(s)$

THINGS TO DO AND THINK ABOUT

1. Jennifer carried out an investigation into energy changes in chemical reactions. She mixed a selection of different chemicals recording the temperature before and after they reacted. Jennifer's results are shown in the table.

Experiment	Reactants mixed	Temperature of A (°C)	Temperature of B (°C)	Temperature of C (°C)	Average temperature of reactants (°C)	Temperature after mixing (°C)
1	A and B	23.4	22.4	-	22.9	31.1
2	A and C	24.6	-	25.0	X	16.4
3	B and C	-	Y	20.4	20.3	26.9

(a) Calculate the temperature change in experiment 1.
(b) Calculate the values of temperatures X and Y.
(c) Jennifer concluded that all the reactions in her experiments were exothermic.
    By referring to the data in the table, explain if Jennifer's conclusion is correct.

2. The table gives information about the physical states of a variety of different chemicals which were used in two different reactions. Use the information in the table to give the state symbols for numbers (i) to (viii).

Substance	State
hydrogen chloride	solution
sodium hydroxide	solution
hydrogen	gas
zinc	solid
aluminium chloride	solution
aluminium hydroxide	solid
sodium chloride	solution
zinc(II) chloride	solution

Reaction 1 $Zn(i) + 2HCl(ii) \rightarrow ZnCl_2(iii) + H_2(iv)$
Reaction 2 $AlCl_3(v) + NaOH(vi) \rightarrow Al(OH)_3(vii) + NaCl(viii)$

3. Write a word equation for each of the following reactions:
   (a) In a blast furnace, carbon monoxide and iron oxide react to produce iron and carbon dioxide.
   (b) Hydrogen gas and magnesium chloride are produced when magnesium reacts with dilute hydrochloric acid.
   (c) When electricity is passed through a solution of copper chloride, chlorine and copper are formed at the electrodes.

4. Write an equation, using symbols and formulae, for each of the following word equations:
   (a) sulfur dioxide + oxygen → sulfur trioxide
   (b) nitrogen + hydrogen → nitrogen hydride
   (c) magnesium + chlorine → magnesium chloride

ACIDS AND BASES

WHAT ARE ACIDS AND BASES?

The term 'acid' comes from a Latin word meaning 'sour'. We encounter many acids in everyday life. Vinegar (literally 'sour wine') is a dilute solution of ethanoic (acetic) acid. Limes, lemons and oranges contain citric acid, tea contains tannic acid, vitamin C is ascorbic acid, and rainwater and fizzy drinks contain carbonic acid. Proteins are built out of long chains of amino acids and even the DNA molecule itself is an acid (deoxyribonucleic acid).

You will also use dilute hydrochloric acid, sulfuric acid and nitric acid in your chemistry experiments in school. Acids used in the food and drink industry can have adverse effects on human health. For example, drinking fruit juices and fizzy drinks with a high acid content can lead to tooth decay and contribute to acid indigestion and heartburn.

Bases can be thought of as the chemical opposites of acids. If a base is dissolved in water, chemists call the resulting solution an **alkali**. Bases are found in many household products. Baking soda contains sodium bicarbonate, indigestion remedies contain calcium carbonate and many drain-cleaning products contain sodium hydroxide, which is commonly known as caustic soda.

Using a pH meter

THE pH SCALE

The **pH scale** is a measure of how **acidic** a solution is. It is a continuous range of numbers from 0 to 14. The pH of a solution can be found by adding pH paper or a pH indicator, such as universal indicator, to the solution and matching the colour of the paper or indicator to the colour on a pH colour chart.

A pH meter can also be used to give a more accurate measure of pH.

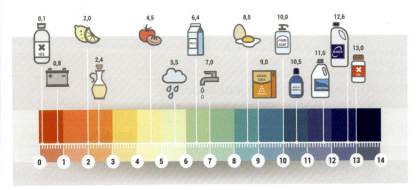

The pH scale

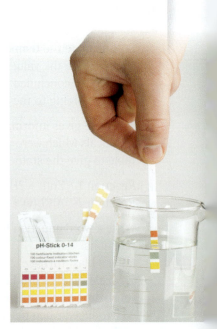

Using pH paper

Any substance with a pH value less than 7 is an acid. The lower the number, the more acidic the substance is.

Any substance with a pH value greater than 7 is an alkali. The higher the number, the more alkaline the substance is.

Pure water and neutral solutions have a pH of exactly 7.

The sulfuric acid in a car battery has a pH less than 1 and is more acidic than vinegar, which has a pH of 3.

Sodium hydroxide solutions can have pH values up to 14 and they are more alkaline than a baking soda solution, which typically has a pH of 8 or 9.

TASK

Do some research to create a PowerPoint presentation or a poster on the uses of acids in the food and drink industries. This should include the names of the acids used and the effects the acids have on human health.

MAKING ACIDS AND ALKALIS

Pure water has a pH of 7 and it turns universal indicator a green colour. If a substance is soluble in water it may alter the pH of the water.

Metal oxides

Metal oxides form when a metal is burned in oxygen.

For example, calcium burning in oxygen will produce calcium oxide. When copper burns in oxygen, copper(II) oxide is produced:

$$\text{calcium} + \text{oxygen} \rightarrow \text{calcium oxide}$$

$$\text{copper} + \text{oxygen} \rightarrow \text{copper(II) oxide}$$

Metal oxides are bases and if they are soluble in water they change the pH to greater than 7.

In this experiment solid calcium oxide powder is added to water containing universal indicator. The calcium oxide dissolves in the water, turning the universal indicator blue and confirming the presence of an alkali.

Copper(II) oxide is a base, but as it does not dissolve in water it cannot alter the pH of the water and so it does not produce an alkaline solution.

Non-metal oxides

Non-metal oxides are produced when non-metals are burned in oxygen.

Non-metal oxides are acidic and if they are soluble in water they produce acidic solutions.

For example, sulfur burning in oxygen will produce sulfur dioxide. When silicon burns in oxygen, silicon dioxide is produced.

$$\text{sulfur} + \text{oxygen} \rightarrow \text{sulfur dioxide}$$

$$\text{silicon} + \text{oxygen} \rightarrow \text{silicon dioxide}$$

In this experiment, yellow sulfur is burned in oxygen gas to produce sulfur dioxide gas. The sulfur dioxide gas formed dissolves in the water, turning the universal indicator red and confirming that an acid has been produced.

Silicon oxide is an acidic oxide, but as it does not dissolve in water it cannot alter the pH of the water and so it does not produce an acidic solution.

DON'T FORGET

A chemical will only change the pH of water if it is soluble in water.

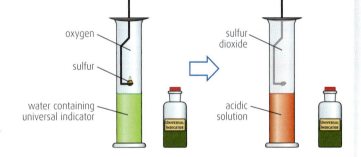

DON'T FORGET

Soluble metal oxides make alkalis when added to water and soluble non-metal oxides make acids when added to water.

THINGS TO DO AND THINK ABOUT

1. Consider the data in the table

Solution	pH
vinegar	3
baking soda	9
sodium chloride	7
lemonade	2

(a) Identify the solutions which are acidic.

(b) Identify the solution with the same pH as water.

(c) Identify the alkaline solution.

NEUTRALISATION

WHAT IS A BASE?

A **base** is a substance which **neutralises** an acid. There are three main types of base: metal oxides, metal hydroxides and metal carbonates. If a base is soluble in water it forms an alkali.

Burning magnesium produces the base magnesium oxide

WHAT IS A NEUTRALISATION REACTION?

When a base is added to an acid, a neutralisation reaction takes place. In a neutralisation reaction the pH of the acid will rise towards 7 as the base is added. If an acid is added to a base the pH will fall towards 7. Water and new substances called **salts** are always formed in a neutralisation reaction.

Naming salts

We can work out which salt is formed in a neutralisation reaction from the name of the acid and base involved in the reaction.

The first part of the name of the base used identifies the first part of the name of the salt. The acid used identifies the second part of the name of the salt.

The base calcium hydroxide is also known as lime

Salts formed from hydrochloric acid are called chlorides, sulfuric acid makes sulfate salts and nitric acid salts are called nitrates.

The table shows some examples of salts produced in neutralisation reactions.

Base	Acid	Salt
sodium hydroxide	hydrochloric	sodium chloride
magnesium oxide	sulfuric	magnesium sulfate
lithium carbonate	nitric	lithium nitrate
potassium hydroxide	sulfuric	potassium sulfate

Word equations for neutralisation reactions

1. Acids and metal oxides

When a metal oxide reacts with an acid, a salt and water are formed: metal oxide + acid → salt + water

> **EXAMPLE**
>
> Copper(II) oxide reacts with dilute sulfuric acid to form copper(II) sulfate and water.
>
> **Word equation:** copper(II) oxide + sulfuric acid → copper(II) sulfate + water

The base strontium carbonate produces the bright red colour in fireworks

2. Acids and alkalis

Alkalis react with acids to form a salt and water:
acid + alkali → salt + water

> **EXAMPLE**
>
> Sodium hydroxide solution reacts with dilute hydrochloric acid forming sodium chloride and water.
>
> **Word equation:** sodium hydroxide + hydrochloric acid → sodium chloride + water

3. Acids and metal carbonates

Metal carbonates react with acids to form a salt, water and carbon dioxide: metal carbonate + acid → salt + water + carbon dioxide

> **EXAMPLE**
>
> Calcium carbonate reacts with dilute nitric acid to form calcium nitrate, water and carbon dioxide.
>
> **Word equation:** calcium carbonate + nitric acid → calcium nitrate + water + carbon dioxide

Following the course of a neutralisation reaction

When preparing a salt in a neutralisation reaction it is important to know when enough base has been added to fully neutralise the acid.

1. Adding an insoluble base

The base is added in small quantities to a flask containing an acid. As the base reacts with the acid it will disappear into the solution forming a salt and water. When the acid has been neutralised any additional base added to the flask will no longer react. It would be obvious that the acid had been neutralised as the excess base would no longer dissolve into the solution and would eventually settle on the bottom of the flask.

Consider the reaction of copper(II) oxide with dilute sulfuric acid:

The unreacted copper(II) oxide on the bottom of the flask indicates that the sulfuric acid has been neutralised.

solid copper(II) oxide

blue copper(II) sulfate solution

sulfuric acid

excess solid copper(II) oxide

1. Adding a soluble base

When a soluble base (alkali) is added to an acid, a pH indicator is used to determine the point at which the acid has been neutralised.

Consider the reaction of sodium hydroxide with hydrochloric acid:

The change in colour of the universal indicator to green means enough sodium hydroxide has been added to neutralise the sulfuric acid.

sulfuric acid with universal indicator

neutral sodium chloride solution

increasing volume of sodium hydroxide added

> **DON'T FORGET**
>
> In a neutralisation reaction a salt and water are always produced. If the base is a carbonate, carbon dioxide will also be formed.

> **DON'T FORGET**
>
> An indicator is not required to show when an acid has been neutralised if the base added to the acid is insoluble in water.

THINGS TO DO AND THINK ABOUT

1. Name the products formed for each of the following neutralisation reactions.

   (a) sodium hydroxide and nitric acid
   (b) lithium carbonate and hydrochloric acid
   (c) potassium carbonate and sulfuric acid
   (d) zinc oxide and hydrochloric acid

2. Which of the following chemicals are bases?

   (a) sodium chloride   (d) magnesium carbonate
   (b) lithium hydroxide   (e) sulfur dioxide
   (c) copper(II) nitrate   (f) potassium sulfate

3. Jill added an alkali to an acid and measured the pH as shown in the diagram.

   The alkali was added until the solution was neutral.

   (a) Suggest the initial pH of the acid solution.
   (b) State the pH when the solution was neutral.
   (c) Name the salt formed in the reaction.
   (d) Write a word equation for the reaction.
   (e) What could have been used instead of a pH meter to determine when the solution was neutral?

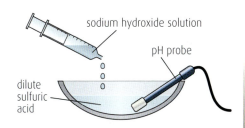

sodium hydroxide solution

pH probe

dilute sulfuric acid

KEY AREA QUESTIONS

RATES OF REACTION

1. Chris added zinc to dilute hydrochloric acid and measured the volume of hydrogen gas produced.

   His results for two experiments at different temperatures are shown below.

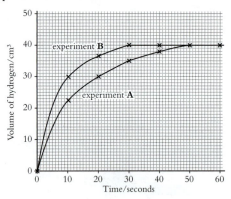

   (a) Suggest a piece of apparatus Chris could have used to measure the volume of hydrogen as the reaction progressed.
   (b) State the total volume of hydrogen gas collected in experiment A?
   (c) State which experiment, A or B, was carried out at the higher temperature.
   (d) Suggest another way to monitor the progress of this reaction.

2. Sally was investigating the reaction between chalk, a form of calcium carbonate, and a dilute acid. Sally carried out the experiment three times, each time using 50 cm³ of the same concentration of sulfuric acid.

   The other reaction conditions Sally used are shown in the table below

Experiment	Mass of chalk (g)	Temperature (°C)
1	1.0	25
2	0.5	35
3	2.0	45

   In each experiment all of the chalk was used up.

   Sally sketched the following graph from the results of experiment 1.

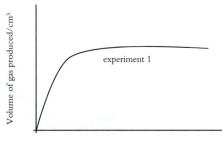

   Copy the graph and add to it the curves you would expect for experiments 2 and 3.

3. Brendan mixed some magnesium with dilute hydrochloric acid. The graph shows how the concentration of hydrochloric acid changes as it reacts with the magnesium.

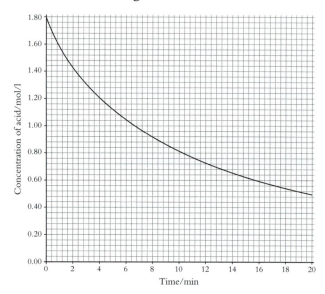

   (a) Why does the graph slope downwards?
   (b) Calculate the change in the concentration of the acid in the first 4 minutes.
   (c) Explain why the graph suggests that the magnesium and dilute hydrochloric acid would still be reacting after 20 minutes.

ATOMIC STRUCTURE AND BONDING RELATED TO PROPERTIES

1. Aluminium oxide is a common ingredient in sunscreen and is sometimes present in cosmetics such as lipstick and nail polish.

   Aluminium oxide is produced when aluminium reacts with oxygen.

   (a) Write a word equation for the formation of aluminium oxide.
   (b) Write the formula for aluminium oxide.
   (c) What type of bonding is present in aluminium oxide?
   (d) The table contains information about an aluminium atom and an oxygen atom.

Atom	Atomic number	Mass number	Number of protons	Number of neutrons
Aluminium	13			14
Oxygen		16	8	

   Copy and complete the table.

2. Hydrocarbons are chemical compounds containing carbon and hydrogen.

Name	Full structural formula	Boiling points (°C)
methane	$\begin{matrix} & H & \\ & | & \\ H- & C & -H \\ & | & \\ & H & \end{matrix}$	–164
ethane	$\begin{matrix} & H & H & \\ & | & | & \\ H- & C & - C & -H \\ & | & | & \\ & H & H & \end{matrix}$	–89
propane	$\begin{matrix} & H & H & H & \\ & | & | & | & \\ H- & C & - C & - C & -H \\ & | & | & | & \\ & H & H & H & \end{matrix}$	–42
butane	$\begin{matrix} & H & H & H & H & \\ & | & | & | & | & \\ H- & C & - C & - C & - C & -H \\ & | & | & | & | & \\ & H & H & H & H & \end{matrix}$	–1
pentane	$\begin{matrix} & H & H & H & H & H & \\ & | & | & | & | & | & \\ H- & C & - C & - C & - C & - C & -H \\ & | & | & | & | & | & \\ & H & H & H & H & H & \end{matrix}$	

(a) What type of bonding is present in a hydrocarbon?

(b) Write the formula for butane.

(c) Calculate the formula mass of propane.

(d) Predict the boiling point of pentane.

(e) Methane is used as a fuel for central heating systems. When this gas burns it reacts with oxygen and produces carbon dioxide and water.

Write a word equation for this reaction.

3. The diagram shows an atom of neon.

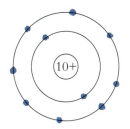

(a) What name is given the central part of an atom?

(b) Why is this atom neutral?

(c) The mass number of this atom is 21. How many neutrons are present in this atom?

ENERGY CHANGES OF CHEMICAL REACTIONS

1. When an alkali is added to an acid, an exothermic reaction takes place.

(a) What will happen to the temperature of the reaction mixture when an alkali is added to an acid?

(b) Which term is used to describe a reaction which takes in energy from the surroundings?

ACIDS AND BASES

1. Calcium oxide and calcium carbonate are used to help remove the acidic gases produced in power stations.

(a) Name the type of reaction that occurs when calcium oxide removes the acidic gases produced in a power station?

(b) State one harmful effect on the environment caused by the release of acidic gases into the atmosphere (see the section on acid rain, p39.)

(c) If some calcium oxide is added to water the pH changes.

   (i) State the pH of pure water.

   (ii) What could be added to the calcium oxide solution to determine the pH?

   (iii) Suggest a pH for the calcium oxide solution.

   (iv) If calcium carbonate is added to water the pH remains unchanged. Suggest a reason for this. You may wish to use the databook.

2. Three common acids used in school laboratories are hydrochloric acid, sulfuric acid and nitric acid.

Kerry was asked to prepare a dry sample of the salt potassium nitrate by adding an acid to a solution of potassium hydroxide.

(a) Name the acid Kerry should use to prepare the salt potassium nitrate.

(b) In addition to the salt, water is also produced in the reaction. Write a word equation for this reaction.

(c) How could a sample of dry potassium nitrate be recovered from the reaction mixture?

NATURE'S CHEMISTRY

FUELS

WHAT IS A FUEL?

A **fuel** is any substance that is burned to produce heat.

There are many different types of fuel. One of the first fuels used by humans was wood.

Like many fuels, the energy stored within wood comes from the Sun. The trees capture the energy from the Sun during photosynthesis. The energy is used to make a chemical called glucose.

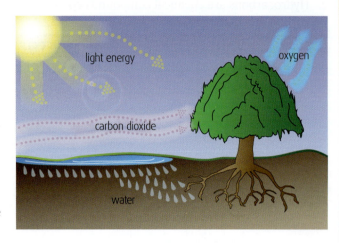

light energy

oxygen

carbon dioxide

water

Some of this energy is turned to heat when the wood is burned

Chemicals in the wood have energy stored in them

Glucose is then used to make other chemicals to allow the tree to grow. One of the main chemicals found in trees is cellulose – a chemical which is made by bonding together thousands of glucose units.

These bonds act as a store for some of the energy that was captured during photosynthesis. When the wood is burned some of the energy stored in the bonds in the chemicals is released as heat.

DON'T FORGET

Combustion is another word for burning. During combustion, the elements in a fuel will react with oxygen.

EXAMPLES OF FUELS

Any fuel has energy from the Sun stored within chemical bonds. There are many examples of these.

🎯 TASK

The picture above shows some of the different types of fuel that we use. There are many different types of fuel and all of them are burned to release some of the chemical energy that is stored inside them. How many different fuels can you name?

FOSSIL FUELS

Many of the fuels shown in the pictures on p.30 were formed from prehistoric living things that existed on the planet around 300 million years ago. These fuels are known as **fossil fuels**. Coal, oil and natural gas are all examples of fossil fuels.

Energy from the Sun was trapped in ancient plants and trees by photosynthesis and the energy was stored in chemical bonds. Ancient animals fed from the plants and trees and so some of the energy was transferred and stored in chemical bonds in these animals. When these ancient plants and animals died, their remains decomposed and eventually turned into fossil fuels.

Fossil fuels are **finite** energy sources. This means that they will one day run out. This is because they are being used far faster than they are being formed.

DON'T FORGET

The word 'fossil' is used for fuels such as coal, oil and natural gas because, like fossils, they are formed from the remains of living things.

Coal

Coal is formed from dead plants. Around 300 million years ago the climate was warmer and so trees and plants grew tall. Sea levels rose and flooded the land causing swamps to form. When the trees and plants died, they fell into these swamps. There was little oxygen present in these swamps and so the trees and plants did not rot. Instead they were eventually covered with layers and layers of mud and sand.

The pressure of these layers of mud and sand, combined with the heat from the Earth's crust, eventually caused the dead plants to decompose and become coal. This whole process took millions of years.

Hundreds of millions of years ago, swamps with giant plants covered the Earth.

Water and soil covered the plant remains 100 million years ago.

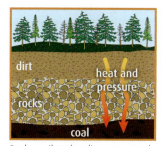

Rocks, soil and sediment created pressure and heat to form coal deep in the ground.

Oil and gas

Oil and gas formed from the remains of dead sea plants and animals. These dead sea plants and animals were buried on the ocean floor and covered with layers and layers of sand and silt. Over millions of years the heat and pressure caused by burial turned them into oil and gas.

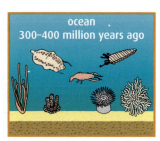

Tiny sea plants and animals died and were buried on the ocean floor. Over time, they were covered by layers of silt and sand.

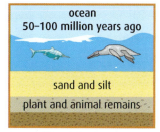

Over millions of years, these remains were buried deeper and deeper. The enormous heat and pressure turned them into oil and gas.

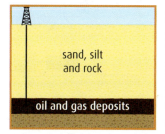

Today, we drill down through layers of sand, silt and rock to reach the rock formations that contain oil and gas deposits.

THINGS TO DO AND THINK ABOUT

Peat is an important substance in Scotland's history. Try to find out some information about peat.

- How is it formed?
- How is it extracted?
- In what way is it similar to coal?
- Is it a fossil fuel?

BURNING OF FOSSIL FUELS

WHAT ARE FOSSIL FUELS MADE OF?

Fossil fuels provide a useful store of energy and are incredibly useful in helping to meet the energy demands of our society. Fossil fuels are mainly made up from compounds called hydrocarbons. It is the hydrocarbons in the fuel that burn to provide the energy. Fossil fuels also contain impurities resulting from their formation underground. Fossil fuels from different areas will contain different impurities. A common impurity in coal is sulfur. Oil as it is extracted from the ground is known as **crude oil** and is made up of many different hydrocarbon compounds. You will learn more about this on pp.36 and 37.

COMBUSTION OF FOSSIL FUELS

When fossil fuels burn, the hydrocarbons react with oxygen in the air. This reaction can also be described as an **oxidation** reaction. Combustion of fuels results in energy being released.

A reaction that releases energy is an **exothermic** reaction.

A reaction that takes in energy is an **endothermic** reaction.

This picture of the whoosh bottle experiment shows just how much energy can be released when a hydrocarbon fuel is burned!

You may have carried out an experiment to test the compounds produced when a hydrocarbon fuel is burned.

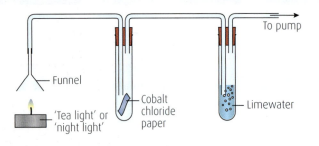

Funnel

'Tea light' or 'night light'

Cobalt chloride paper

Limewater

To pump

DON'T FORGET

Combustion is another word for burning. Combustion is an oxidation reaction. Combustion is an exothermic reaction.

Indicator	Observations
Cobalt chloride paper	Turned pink
Limewater	Turned cloudy

Before burning the hydrocarbon fuel, the equipment is attached to a pump which pulls the gases, produced during combustion, through the equipment.

Blue cobalt chloride paper turns pink when water is present.

Limewater turns cloudy when carbon dioxide gas is present.

The results in the table show that both water and carbon dioxide are produced when a hydrocarbon fuel burns. In fact, carbon dioxide and water are produced when *any* hydrocarbon burns in a plentiful supply of oxygen.

A word equation can be written for this reaction:

hydrocarbon + oxygen → carbon dioxide + water

Incomplete combustion

When there is not enough oxygen present for complete combustion to take place, **incomplete combustion** occurs. This can result in the poisonous gas carbon monoxide being formed:

hydrocarbon + limited oxygen →
carbon monoxide + carbon + water.

Carbon monoxide is an odourless, colourless gas and is highly poisonous. Faulty gas appliances, such as gas boilers, can produce carbon monoxide and, for this reason, gas appliances should be inspected regularly. The fuel burned in a car engine does not get enough oxygen for complete combustion and so carbon monoxide gas is produced. **Catalytic converters** are fitted to cars in order to convert the carbon monoxide gas into safer carbon dioxide gas.

If there is even less oxygen present when a hydrocarbon burns, a black soot will be formed. Soot can be seen on the bottom of a beaker of water if it has been heated using the yellow flame of a Bunsen burner.

Soot forming on the bottom of a beaker when heated with a Bunsen burner.

THINGS TO DO AND THINK ABOUT

1. When any substance burns, the elements within that substance react with oxygen to form oxides. When hydrogen burns it combines with oxygen to make hydrogen oxide (water). When carbon burns it produces carbon dioxide. Write word equations to show the reaction that takes place when the substances below are burned completely in oxygen.

   - Magnesium
   - Iron
   - Sulfur
   - Carbon hydride

2. What is the definition of the following words?

   - Combustion
   - Oxidation
   - Exothermic
   - Endothermic

COMBUSTION

From the previous section we know that when a substance burns it reacts with oxygen. We are now going to look at combustion reactions in more detail.

CONSERVATION OF MASS

The idea of conservation of mass was first proposed by Antoine Lavoisier in the 18th century.

He discovered that when a substance combusts, mass is conserved. In other words, the total mass of all the products at the end of a reaction is the same as the total mass of all the reactants.

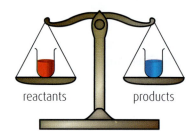

reactants   products

Antoine Lavoisier (1743–1794).

EXAMPLE

You may have seen an experiment such as that shown in the diagram below:

iron wool   plasticine

Before the experiment   After the experiment

The experiment is set up so that the plasticine is slightly heavier than the iron wool. The iron wool is set alight using a Bunsen burner. The combustion reaction causes the iron wool to react with oxygen in the air. This means that the iron atoms combine with oxygen atoms in the air. Many people think that this will make the product lighter, but actually the product is heavier than the iron wool reactant.

The product of the reaction is a substance containing the same number of iron atoms as the reactants, but these atoms are now joined with oxygen atoms, which also have mass. If we had also measured the mass of the oxygen that was used in the experiment, we would have seen that the total mass of the reactants and products was equal.

What is the name of the compound made during this reaction?

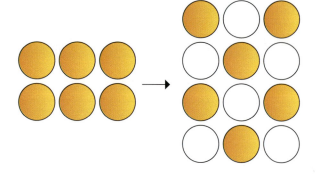

Atoms in iron wool   Atoms in product of burning
before burning   iron wool in oxygen

TASK

Lavoisier was responsible for identifying oxygen. Before this time, scientists believed in another theory to help explain combustion. Try to find out the name of this theory.

COMBUSTION OF NATURAL GAS

Conservation of mass applies to all reactions, including the combustion of hydrocarbon fuels. In this case, one of the products is a gas and this gas is usually left to escape into the atmosphere so the products *appear* to have less mass than the reactants.

Can you remember the name of the gas produced when a hydrocarbon burns?

Natural gas is mainly made up of a hydrocarbon compound called methane. This is the simplest hydrocarbon compound. Like all hydrocarbons, when it burns in a plentiful supply of oxygen, carbon dioxide gas and water are produced. The word equation for this reaction is:

$$\text{methane} + \text{oxygen} \rightarrow \text{carbon dioxide} + \text{water}$$

The total mass of methane and oxygen is equal to the mass of carbon dioxide and water produced.

FIRE TRIANGLE

Another discovery of Antoine Lavoisier was that oxygen was essential for combustion to take place. Without oxygen, substances do not burn. In fact, there are two other essential ingredients in a combustion reaction – fuel and heat. These three essential ingredients can be represented in a triangle.

If any of these three ingredients are removed then combustion stops. It is this information that is used to put out fires.

Removing fuel

An example of removing fuel from a combustion reaction is switching off the gas tap when using a Bunsen burner.

Removing heat

Water is used to extinguish some types of fire. By putting water onto a fire, heat is being removed from the combustion reaction. However, water cannot be used on all types of fires. For example, if water was used on a hydrocarbon fuel fire, the water would spread the fire instead of putting it out.

Removing oxygen

Fire blankets and carbon dioxide fire extinguishers act by reducing the amount of oxygen a burning fuel can obtain. When a fire blanket is put onto a fire it smothers the fire and stops oxygen from coming into contact with the fuel. Carbon dioxide works in a similar way. It is more dense than oxygen (air) and so sinks and covers the fire in a blanket-like layer of carbon dioxide gas.

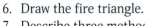

THINGS TO DO AND THINK ABOUT

1. Lavoisier proposed the law of conservation of mass. What does this mean?
2. What happens to the mass of iron wool when it is burned? Explain your answer.
3. Why do some combustion reactions appear to have products that have less mass than the reactants?
4. Name the main hydrocarbon in natural gas?
5. Write a word equation showing the compound in natural gas burning completely.
6. Draw the fire triangle.
7. Describe three methods that could be used to put out a fire.

Natural gas is the gas that supplies many homes in Scotland and is used in domestic cooking and heating.

CRUDE OIL

The oil that is extracted from oil wells, often under the sea, is known as **crude oil**. This is a black, thick, sticky substance (viscous) and is made up of many different hydrocarbon molecules.

Many important products come from the compounds in crude oil. As well as fuels such as petrol and diesel, compounds in crude oil are used in making plastics, dyes and medicines, to name only a few. Crude oil itself, though, is not particularly useful. It is not used as a fuel, for example, as it does not burn well. To obtain the useful compounds and fuels contained in crude oil it is first necessary to separate the mixture of hydrocarbons contained within it.

FRACTIONAL DISTILLATION

Distillation is a common technique used to separate a mixture when at least one of the substances in the mixture is a liquid. It makes use of the different boiling points of the different substances in the mixture and results in the original mixture being separated into pure substances.

DON'T FORGET

Something which is viscous has a thick, sticky consistency, between that of a liquid and a solid. Another example of a viscous substance is syrup.

TASK

How many different methods to separate mixtures can you think of?

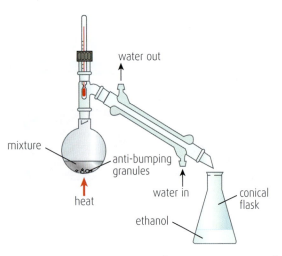

Diagram of distillation equipment used to separate a mixture of ethanol and water.

Crude oil is a **viscous** liquid containing many different compounds. To separate the mixture a process called **fractional distillation** is used. This is carried out in an oil refinery, like the one at Grangemouth in central Scotland.

Fractional distillation is similar – a mixture of substances is separated by making use of the different boiling points of the substances present. It differs only that the end result is not a separation into pure substances, but a separation into different parts or fractions. Each fraction still contains a mixture of different compounds.

Have a look at the diagram on the right of a typical distillation still used to separate crude oil.

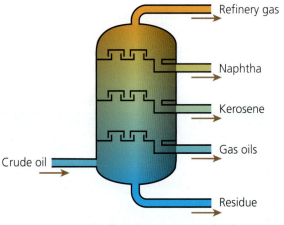

Typical still used to separate crude oil

TASK

Try to find the names of three types of chemicals that can be made using the naphtha fraction of crude oil.

Crude oil is heated and then poured into the bottom of a long column fitted with a number of condensers at different heights. The column is hottest at the bottom and coolest at the top. The gases in the crude oil mixture rise up through the column and, as the temperature of the column lowers, some parts of the mixture **condense** back into a liquid again. The parts of the mixture with the highest boiling points are found near the bottom of the column (residue); those with low boiling points will condense at the top of the column (refinery gas). These different parts of the mixture are collected and are called **fractions**.

DON'T FORGET

A **fraction** is a group of hydrocarbon molecules that all have boiling points within a certain range.

PROPERTIES OF THE FRACTIONS

As you have already seen, different fractions have different boiling points. Each fraction contains a number of different hydrocarbon molecules. Fractions containing small hydrocarbon molecules have low boiling points and fractions containing large hydrocarbon molecules have high boiling points.

Fraction	Temperature range (°C)	Approximate number of carbon atoms per molecule
Refinery gas	<30	1–4
Petrol	30–120	5–9
Naphtha	60–130	6–11
Kerosene	130–240	11–16
Diesel (gas) oil	240–350	15–25
Residue	>350	>25

The table on the right shows typical boiling point ranges for fractions obtained when distilling crude oil.

The size of the hydrocarbon molecules in each fraction has an effect on other properties too. Fractions containing small molecules **evaporate** easier and are more **flammable**. The **viscosity** of the fractions also changes. Fractions containing small molecules are less viscous than those containing large molecules. The table below summarises the properties of the fractions

Fraction	Number of carbon atoms per molecule	Boiling point	Ease of evaporation	Viscosity	Flammability
Refinery gas	Small	Low	Easy	Low	High
Petrol					
Naphtha					
Kerosene					
Diesel (gas) oil					
Residue	Large	High	Difficult	High	Low

THINGS TO DO AND THINK ABOUT

1. Crude oil is a **viscous** substance. What is the meaning of **viscous**?
2. What does crude oil contain?
3. Name three important products that are obtained from crude oil.
4. What process is used to separate the different compounds in crude oil?
5. What name is given to the different parts that crude oil is separated into?
6. Name the six different parts formed when crude oil is separated.
7. What property of the different parts of crude oil allows it to be separated?
8. Which part has the smallest molecules?
9. How does ease of evaporation change as you move from the smallest molecules to the biggest molecules?
10. Which part is the most viscous?

IMPACT ON THE ENVIRONMENT

We have seen that products from crude oil as well as coal and natural gas are all very important in modern life. You may have also heard about other sources of energy such as wind and solar. Why is the development of alternative energy sources necessary? To explain this we need to know a little about the carbon cycle.

CARBON CYCLE

Carbon is present in all living things. When these living things die, the carbon they contain is recycled by other living things and so it is in a cycle.

There are four main stages to the carbon cycle:

1. Carbon dioxide in the atmosphere is absorbed by plants to make carbohydrates.
2. Animals eat the plants and so pass the carbon along the food chain. Most of the carbon they eat is given out as carbon dioxide when the animal breathes.
3. The plants and animals die and are decomposed. This releases carbon dioxide back into the environment.
4. Carbon dioxide from burning fossil fuels also contributes to the cycle.

This cycle ensures that the levels of carbon dioxide in the atmosphere are kept constant.

Fossil fuels are an ancient source of carbon. They formed because some dead plants and animals did not decompose and so the carbon was stored underground rather than being released back into the carbon cycle as carbon dioxide.

We have already seen that coal, oil and natural gas were formed millions of years ago. At this time carbon dioxide levels were higher and temperatures were warmer. The process that formed them took a very long time. Fossil fuels are being used up far faster than they are being formed and so they are a **finite energy source**, in other words, they will run out.

TASK

As well as fuels, other products such as plastics, dyes, medicines, fertilisers (used to provide large quantities of food) and lots more important chemicals are made from fossil fuels. Try to think how your day today would have been different if there were no fossil fuels.

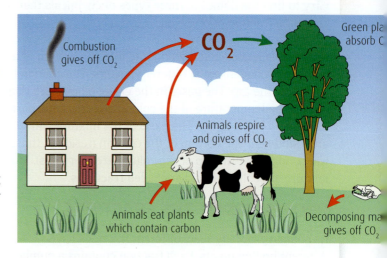

Combustion gives off CO_2

CO_2

Green pla absorb C

Animals respire and gives off CO_2

Animals eat plants which contain carbon

Decomposing ma gives off CO_2

GREENHOUSE EFFECT AND GLOBAL WARMING

You may have heard the terms **greenhouse effect** and **global warming** before – what do you think these two terms mean?

Greenhouse effect

Without the greenhouse effect the Earth would be a very cold place. Greenhouse gases such as carbon dioxide and methane form a layer, or blanket, around the planet. This traps some of the energy that is given out by the surface of the Earth, keeping the planet warm. Without these greenhouse gases, much more energy would escape and the overall temperature would be much cooler.

DON'T FORGET

The greenhouse effect is a natural process that helps to keep the planet from being too cold to sustain life.

Global warming

Global warming is also known as climate change and refers to unpredictable and rapidly changing weather and temperatures that affect the whole planet. Global warming is mainly caused by the increased quantity of greenhouse gases in the atmosphere. This causes more heat to be trapped, leading to warmer temperatures. There are many reasons for the increase in the quantity of greenhouse gases, but much of the increase is due to human activities such as burning fossil fuels.

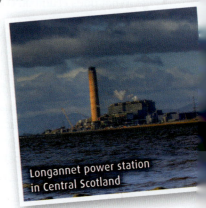

Longannet power station in Central Scotland

The greenhouse effect

Earth's surface is heated by the Sun and radiates the heat back out towards space

Some enrgy is reflected back out into space

Greenhouse gases in the atmosphere trap some of the heat

Solar energy from the Sun passes through the atmosphere

Burning fossil fuels releases large quantities of carbon dioxide into the atmosphere at a rate that is too fast for natural processes in the carbon cycle to be able to reabsorb it. Carbon dioxide levels in the atmosphere are therefore increasing.

Although the effects of global warming cannot be accurately predicted, some scientists believe there could be more floods in some parts of the world, droughts in others, melting of ice caps, rising of sea levels as well as many other effects.

DON'T FORGET

Global warming is caused by increasing the quantity of greenhouse gases (such as carbon dioxide and methane) in the atmosphere leading to warmer temperatures.

ACID RAIN

Acid rain is rain that has higher than normal quantities of acid in it, which means that acid rain has a lower pH value than 'normal' rain.

Acid rain is formed when gases such as nitrogen oxides and sulfur dioxide dissolve in rain water to produce nitric acid and sulfurous acid.

Nitrogen oxides are produced in the atmosphere naturally during lightning. The energy formed from lightning causes the nitrogen and oxygen in the air to react together, producing nitrogen oxides. The main human source of nitrogen oxides in the atmosphere is from petrol engines, where the sparks of the spark plugs (needed to ignite the petrol–oxygen mixture) also cause nitrogen and oxygen from the air to react together.

Sulfur dioxide is produced naturally when volcanoes erupt but the biggest human source of sulfur dioxides is from burning fossil fuels, particularly in factories and fossil fuel power stations. Sulfur impurities in fossil fuels react with oxygen when fossil fuels burn and this leads to the formation of sulfur dioxide:

sulfur + oxygen → sulfur dioxide

Acid rain causes widespread problems. It can have devastating effects on plants and animals in rivers and lakes. Forests in some parts of the world have been badly affected by acid rain. Rocks and cliff faces are also affected as some types of rock will react with acid. Acid rain has also been shown to damage cars, buildings, statues, rail tracks, bridges and many more things.

DON'T FORGET

There are a few different compounds containing nitrogen and oxygen and these are referred to simply as 'nitrogen oxides' or 'oxides of nitrogen'.

THINGS TO DO AND THINK ABOUT

1. What is the **greenhouse effect**?
2. What would happen to the Earth if there were no greenhouse gases?
3. Name two greenhouse gases.
4. What is **global warming**?
5. What is thought to be causing global warming?
6. What is acid rain?
7. Name one gas that causes acid rain and write down how it is formed.
8. Name the other gas that dissolves in rain water producing acid rain?
9. How is this gas formed?
10. List some of the harmful effects of acid rain.

PROTECTING THE ENVIRONMENT

In the previous section we learned about some of the disadvantages of using fossil fuels.

 TASK

Can you list five ways in which fossil fuels are bad for the environment?

The most obvious way to prevent some of these negative effects on the environment is to reduce our use of fossil fuels simply by reducing our energy requirements. Examples include: walking, cycling or taking public transport instead of taking the car; switching off lights and other electrical appliances when they are not in use; putting on an extra jumper and turning down the heating. Much research and resources are being put into developing alternative forms of energy as well as looking at ways that the harmful effects of using fossil fuels can be reduced. Some of these are discussed below.

ALTERNATIVE SOURCES OF ENERGY

There are many alternative sources of energy in use in Scotland.

These alternative sources of energy help to reduce the amount of electricity being generated by burning fossil fuels and so help to reduce pollution.

Whitelee wind farm in East Renfrewshire

 TASK

What effects do you think wind, photovoltaic cells and hydroelectric power have on the carbon cycle? As well as the actual pollution you will also need to think of the effect on the carbon cycle of making and building the technology needed for these alternative energy sources.

HYDROGEN AS A FUEL

Research scientists are looking at using hydrogen as an alternative fuel to power vehicles. There are two main ways in which this can be achieved.

Hydrogen combustion

Hydrogen can be used in some vehicles in the same way that petrol is used. The hydrogen fuel is burned in an internal combustion engine.

Photovoltaic cells (solar cells)

There are still some disadvantages to using hydrogen as a fuel that need to be overcome. Hydrogen does not produce as much energy as petrol and diesel per unit volume and so vehicles will not travel as far before the fuel tank needs to be refuelled. There are few filling stations supplying hydrogen and cars that can run on hydrogen are expensive to buy.

Hydrogen fuel cells

In a hydrogen fuel cell, hydrogen and oxygen react to produce water. Electricity is also produced. Fuel cells are often regarded as batteries because they produce electricity from a chemical reaction – the main difference is that hydrogen fuel cells do not run out of charge provided the hydrogen fuel is constantly supplied.

The dam at Cruachan Hydroelectric Power Station in Argyll

In a space shuttle, hydrogen fuel cells also power the shuttle's electrical systems, producing a clean by-product, pure water, which the crew can then drink.

TASK

Write a word equation for the combustion of hydrogen.

Did you write hydrogen oxide as the product of burning hydrogen? The more common name for hydrogen oxide is water. Burning hydrogen has one main advantage over fossil fuels – the product, water, is much safer for the environment.

TASK

Find out the main method used to make hydrogen. What disadvantages are there to this method of producing hydrogen? Can hydrogen really be regarded as a 'clean' fuel?

OTHER TECHNOLOGIES

Scientists are also looking at other methods of reducing the amount of carbon dioxide in the atmosphere. One current area of development in Scotland is carbon capture and storage. This involves first capturing the carbon dioxide produced by factories and power stations. This carbon dioxide is then turned into a liquid and stored in old oil and gas wells deep under the ocean.

A hydrogen filling station

CATALYTIC CONVERTERS

So far we have learned about methods to reduce the quantity of fossil fuels being used as well as learning about methods to reduce the level of carbon dioxide in the atmosphere, but what about the other pollutants from burning fossil fuels?

We have seen that petrol engines produce nitrogen oxides. This is one of the gases that contributes to acid rain. Another pollutant from burning fossil fuels is carbon monoxide. This is produced when there is not enough oxygen to allow the fuel to burn completely.

Catalytic converters are fitted to all modern cars to help to reduce all these pollutants. Three metals, platinum, palladium and rhodium, are used in catalytic converters and these metals catalyse the reaction of nitrogen oxides and carbon monoxide into nitrogen and carbon dioxide.

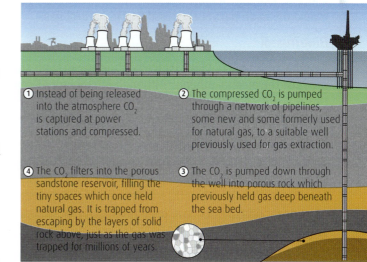

① Instead of being released into the atmosphere CO_2 is captured at power stations and compressed.

② The compressed CO_2 is pumped through a network of pipelines, some new and some formerly used for natural gas, to a suitable well previously used for gas extraction.

③ The CO_2 is pumped down through the well into porous rock which previously held gas deep beneath the sea bed.

④ The CO_2 filters into the porous sandstone reservoir, filling the tiny spaces which once held natural gas. It is trapped from escaping by the layers of solid rock above, just as the gas was trapped for miillions of years.

THINGS TO DO AND THINK ABOUT

1. Name three alternative sources of energy.
2. What other methods have scientists developed for reducing the amount of carbon dioxide in the atmosphere?
3. What has been done to reduce the amount of acid rain in the environment?

DON'T FORGET

Combustion is the chemical word for burning. When a substance burns, it combines with oxygen.

DON'T FORGET

As well as contributing to global warming, the burning of fossil fuels can lead to acid rain production. Petrol engines cause nitrogen oxides to be produced and sulfur impurities in fossil fuels lead to sulfur dioxide being produced. Both of these gases dissolve in rain to make it more acidic.

BIOMASS AND BIOFUELS

Biomass fuels are becoming an increasingly popular alternative energy source for both domestic and industrial uses.

Biomass is the name given to substances that are derived from living or recently living organisms. Biomass is often used to refer to plants or plant-based substances.

ENERGY FROM BIOMASS

Biomass can be burned directly to produce energy. The most common example of a biomass fuel is wood.

Other biomass energy sources that are burned directly include some crops as well as rubbish.

Burning wood.

BIOFUELS

Biofuels are fuels that have been made from biomass sources.

Biomethane

Biomethane (biogas) is a naturally occurring gas which is produced by the digestion (in the absence of oxygen) of organic material such as dead animal and plant material, manure, sewage and organic waste. The gas produced is methane. Methane is also the main part of the fossil fuel in natural gas.

Bioethanol

Bioethanol is mainly made by fermenting sugar. High sugar crops are fermented to produce the alcohol ethanol. Although ethanol for alcoholic drinks is also produced this way, the largest use of ethanol produced by fermentation is as a fuel. Bioethanol can be used as fuel for motor vehicles. This idea is not a new one. Henry Ford designed his Model T to run on ethanol.

Bioethanol is usually combined with another fuel such as petrol. Most petrol cars could run on petrol containing 5% ethanol. In the USA, most cars are powered by petrol that contains 10% ethanol. To use a higher percentage of ethanol, engines must be modified. Cars have been produced that can run on as much as 85% ethanol.

Brazil has the highest proportion of road vehicles designed to run on biofuels. This is due to the vast quantities of sugar cane that can be grown in Brazil, providing a large source of sugar that can be fermented.

Model T Ford

Sugar cane

Biodiesel

Biodiesel is an alternative fuel and is similar to 'normal' diesel. Biodiesel can be produced from vegetable oil, animal oil/fats and waste cooking oil. Crops such as rapeseed, palm and soyabean are used to produce oils for biodiesel. In the UK rapeseed is the most widely used oil in biodiesel manufacture.

Biodiesel can also be produced from waste oil from restaurants, chip shops and other food industries.

Like bioethanol, biodiesel is usually mixed with fossil fuel derived diesel before being used as a vehicle fuel. A blend of 10% biodiesel can run in most diesel engines.

Rapeseed crop

COMPARING BIOMASS FUELS WITH FOSSIL FUELS

There are two main ways in which biomass fuels differ from fossil fuels.

1. Biomass based fuels and biofuels are made from sustainably grown crops. This makes them renewable fuels as the crops can be regrown. This gives them an advantage over non-renewable fossil fuels which take millions of years to form.

2. Biomass and biofuels are sometimes regarded as being carbon neutral. Like fossil fuels, burning biomass fuels also produces carbon dioxide. So why can it be described as being carbon neutral? The crops that are grown to produce biomass fuels use carbon dioxide during photosynthesis. This provides the plant with food and energy to grow and so the production and use of biomass fuels follows the carbon cycle and little extra carbon dioxide enters the atmosphere. This cannot be said for fossil fuels. The carbon stored in fossil fuels is from millions of years ago. Releasing this, through burning, adds extra carbon dioxide into the atmosphere.

DON'T FORGET

Biomass based fuels are renewable. They can be described as being carbon neutral as the carbon dioxide produced by burning biomass fuels is absorbed by the plants being grown to produce the fuel. Fossil fuels are non-renewable and are not carbon neutral.

TASK

Burning of biomass fuels leads to less carbon dioxide being released into the atmosphere than burning fossil fuels. Can you think of reasons why they are not fully 'carbon neutral'? You may wish to discuss this with a partner.

THINGS TO DO AND THINK ABOUT

1. What does **biomass** mean?
2. Name a common biomass energy source.
3. What is a biofuel?
4. Copy and complete the table below.

Biofuel	Source	Uses
Biomethane	Rotting rubbish	
Bioethanol		
Biodiesel		

5. Why can biomass fuels be described as carbon neutral?

HYDROCARBON COMPOUNDS - ALKANES

HYDROCARBONS

Hydrocarbons are compounds that contain carbon and hydrogen only. They can be obtained by **fractional distillation** of crude oil and are the main compounds present in **fossil fuels** such as **natural gas** and **oil**.

Hydrocarbons can be divided into different subsets, sometimes called families. This course considers two of these families – alkanes and alkenes.

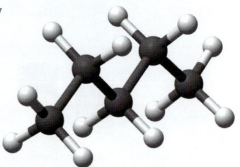

Three-dimensional model of butane.

ALKANE FAMILY

Most of the compounds found in crude oil based fuels such as petrol belong to the **alkane** family. One member of this family is butane. You may have heard of butane before as it is a common fuel used for portable heaters, barbecues and in some camping stoves.

In three dimensions butane can be represented as shown above.

The grey spheres represent carbon atoms and the white spheres represent hydrogen atoms. This means that butane has four carbon atoms and ten hydrogen atoms, and so its **molecular formula** can be written C_4H_{10}.

All of the carbon atoms are joined together by single covalent bonds and so butane is a **saturated hydrocarbon**.

DIFFERENT TYPES OF FORMULA

As well as representing the three-dimensional structure of alkanes as shown above, there are other types of formula that can be used to represent hydrocarbons.

Molecular formula

This is the simplest representation of the compound and shows the symbol for each type of atom present and the number of each type of atom.

The molecular formula for butane is C_4H_{10}.

Full structural formula

The **full structural formula** for butane is:

This shows all the atoms and all the bonds in the molecule.

General formula

The general formula for the alkane family is C_nH_{2n+2} where n represents the number of carbon atoms. This formula can be used to work out the formula of any alkane provided the number of carbon atoms is known. For example, the alkane octane contains eight carbon atoms so, using the general formula, the molecular formula for octane can be calculated as follows:

$$C_8H_{(2\times8)+2} \rightarrow C_8H_{18}$$

Prefix	Number of carbons
meth-	1
eth-	2
prop-	3
but-	4
pent-	5
hex-	6
hept-	7
oct-	8

DON'T FORGET

It is useful to know these prefixes and their meanings but you can also find the names of the first eight alkanes listed in numerical order of carbon atoms in the data book.

NAMING STRAIGHT-CHAIN ALKANES

All hydrocarbons can be named using a systematic naming system. In this system a **prefix** is used that lets other chemists know how many carbons are in the compound. The prefix for butane is 'but-' and this means four.

The ending for an alkane name is 'ane'. This tells a chemist that there are carbon-to-carbon single bonds in the compound. Therefore a hydrocarbon compound can be recognised as an alkane if the name ends with **ane**. An alkane containing three carbon atoms, for example, would be called propane (prop = 3, ane = alkane).

CRACKING

Of the fractions that are distilled from crude oil, the petrol fraction is the one that is in greatest demand. Fractional distillation of crude oil does not allow this demand to be met. Longer chain, less useful hydrocarbon molecules from crude oil are 'split' or **cracked** to make them into smaller, more useful compounds that will help to meet the demand. The process is known as **cracking**.

Cracking is a chemical reaction in which some of the bonds in an alkane molecule are broken to produce smaller molecules. In the laboratory, this experiment can be carried out as shown in the diagram.

High temperatures and a catalyst are needed for this reaction. Aluminium oxide is often used as a catalyst.

As well as producing shorter chain alkane molecules, this reaction also produces molecules known as **alkenes**. A general word equation for the reaction is

alkanes → smaller alkanes + smaller alkenes

Alkane	Full structural formula	Molecular formula
methane		CH_4
ethane		C_2H_6
propane		C_3H_8
butane		C_4H_{10}

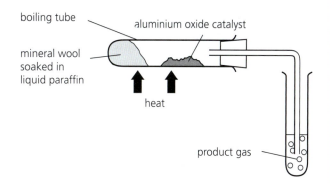

THINGS TO DO AND THINK ABOUT

1. Name a source of hydrocarbon compounds.
2. What does 'saturated' mean when used to describe a hydrocarbon?
3. Explain the difference between a molecular formula and a full structural formula.
4. What prefix is used in naming a hydrocarbon compound containing five carbon atoms?
5. Name the alkane with three carbon atoms joined in a straight chain.
6. Name the following hydrocarbon:

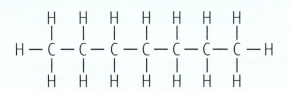

7. State the definition of 'chemical cracking'.
8. What catalyst is used in cracking a long chain hydrocarbon?
9. Name the two types of hydrocarbon compounds formed when an alkane is cracked.
10. Write a general word equation for a cracking reaction.

HYDROCARBON COMPOUNDS - ALKENES

ALKENES: AN INTRODUCTION

Alkenes are another family of hydrocarbons. They are obtained by **cracking crude oil** fractions and can be used to make ethanol and polymers (plastics) – two very important products in modern society.

Alkenes differ from alkanes in that they contain a carbon-to-carbon double covalent bond (C=C) and so are **unsaturated** molecules.

ALKENE FAMILY

Take a look at the table on the right, which gives the formulae of the first four members of the alkene family.

 TASK

Can you work out a general formula for the alkene family? Remember that the letter *n* is used to represent the number of carbon atoms in the molecule.

STRAIGHT-CHAIN ALKENES

In the table on the right the carbon-to-carbon double bond is positioned at the end of the chain. For butene and pentene the C=C could be positioned between different carbon atoms.

If we look at pentene we can see that the C=C could be in two different positions:

Alkene	Full structural formula	Molecular formula
ethene	$\begin{array}{c} H \qquad\qquad H \\ \backslash \qquad\qquad / \\ C = C \\ / \qquad\qquad \backslash \\ H \qquad\qquad H \end{array}$	C_2H_4
propene	$\begin{array}{c} \qquad H \qquad\qquad H \\ \qquad \| \qquad\qquad / \\ H-C-C=C \\ \qquad \| \qquad\qquad \backslash \\ \qquad H \quad H \qquad H \end{array}$	C_3H_6
butene	$\begin{array}{c} \quad H \quad H \qquad\quad H \\ \quad \| \quad \| \qquad\quad / \\ H-C-C-C=C \\ \quad \| \quad \| \quad \| \quad\; \backslash \\ \quad H \quad H \quad H \quad\; H \end{array}$	C_4H_8
pentene	$\begin{array}{c} \quad\; H \quad H \quad H \qquad\quad H \\ \quad\; \| \quad \| \quad \| \qquad\quad / \\ H-C-C-C-C=C \\ \quad\; \| \quad \| \quad \| \quad \| \quad\; \backslash \\ \quad\; H \quad H \quad H \quad H \quad\; H \end{array}$	C_5H_{10}

A   and   **B**

$$\begin{array}{cc} \begin{array}{c} \quad H \quad H \qquad\quad H \\ \quad \| \quad \| \qquad\quad / \\ H-C-C-C=C \\ \quad \| \quad \| \quad \| \quad\; \backslash \\ \quad H \quad H \quad H \quad\; H \end{array} & \begin{array}{c} \quad H \quad H \qquad\qquad H \\ \quad \| \quad \| \qquad\qquad \| \\ H-C-C-C=C-C-H \\ \quad \| \quad \| \qquad\qquad \| \\ \quad H \quad H \quad H \quad H \quad H \end{array} \end{array}$$

Naming straight-chain alkenes

The two structures of pentene shown above have slightly different properties. The boiling point of **A** is 30°C and the boiling point of **B** is 36°C, so they cannot both be called pentene.

To differentiate the two molecules by name, an internationally agreed set of rules is used:

- The number of carbon atoms in the chain is counted and this will determine the prefix (five carbons = pent-).
- The carbon atoms in the chain are numbered, beginning with the end carbon nearest to the double bond. The molecules **A** and **B** would be numbered as shown below.

A $C^5-C^4-C^3-C^2=C^1$ and **B** $C^5-C^4-C^3=C^2-C^1$

- The position of the C=C bond is given by the first carbon in the double bond. In molecule **A** this is carbon 1 and in molecule **B** this is carbon 2.
- The number is inserted into the name using hyphens (-) to separate the words and numbers.
- The ending '-ene' is added. This shows that the compound is a member of the homologous series, the alkenes, and so the compound contains a C=C double bond. **A** is therefore given the name pent-1-ene and **B** is given the name pent-2-ene.

DON'T FORGET

Saturated hydrocarbons like alkanes have carbon-to-carbon single bonds. Alkenes are unsaturated hydrocarbons as they contain a carbon-to-carbon double bond.

TESTING FOR ALKENES

Alkanes and alkenes have different chemical properties. In other words they react differently. This is due to the difference in their structures. Alkanes contain only carbon-to-carbon single bonds and so are **saturated** molecules. Alkenes contain carbon-to-carbon double bonds and so are **unsaturated** molecules. The presence of carbon-to-carbon double bonds make alkene molecules more reactive than alkanes. For example, alkenes will react with bromine solution.

TESTING FOR UNSATURATION

If you have carried out a cracking reaction in class you probably tested the products for unsaturation. To do this, a few drops of bromine water (bromine solution) are added to the products from the cracking reaction. The yellow–orange bromine solution rapidly turns colourless when an unsaturated compound, such as an alkene, is present. This reaction does not take place if an alkane molecule is used. The bromine water does not react with alkanes and so the bromine water remains a yellow–orange colour.

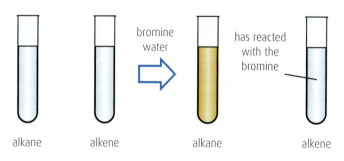

Compound	Reaction with bromine
Saturated	Remains yellow–orange
Unsaturated	Rapidly decolourises

THINGS TO DO AND THINK ABOUT

1. How are alkenes produced?
2. Name two important products made from alkenes.
3. How do alkenes differ from alkanes?
4. What prefix is used to name an alkene containing four carbon atoms joined in a straight chain?
5. What does 'unsaturated' mean when used to describe a hydrocarbon compound?
6. Name the alkene containing seven carbon atoms joined in a straight chain.
7. What is the name of this compound?

8. What is the name of this compound?

9. Describe a test that could be carried out to show that an unknown compound is an alkene.
10. What would be seen when:
    (a) butane was mixed with bromine water?
    (b) butene was mixed with bromine water?

EVERYDAY CONSUMER PRODUCTS

Hydrocarbon compounds are present in many of the consumer products we use on a daily basis, from shampoos and cleaning products to medicines and the food we eat.

PLANTS

We have already seen that plants can capture energy from the Sun and turn this into glucose. This is known as photosynthesis and the word equation for the overall reaction is:

carbon dioxide + water → glucose + oxygen.

Plants use the glucose produced as energy to grow. To store the energy, glucose is converted into starch. Unused energy is converted into oils. Plants also provide many other important chemicals – many of which are used in consumer products.

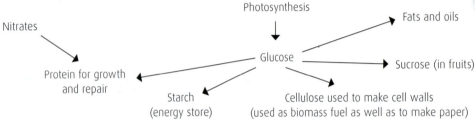

Photosynthesis

Nitrates

Protein for growth and repair

Glucose

Fats and oils

Sucrose (in fruits)

Starch (energy store)

Cellulose used to make cell walls (used as biomass fuel as well as to make paper)

Products of photosynthesis

Consumer products from plants: cotton, aloe vera and paper.

⊙ **TASK**

Try to find the names of five different products that are made from plants.

CARBOHYDRATES

A **carbohydrate** is a compound containing carbon, hydrogen and oxygen. Carbohydrates are essential in our diet and can be split into two groups – sugars and starch. Carbohydrates are a source of energy.

The simplest carbohydrate is **glucose**. It has the formula $C_6H_{12}O_6$. You can see from this formula that there are twice as many hydrogen atoms as oxygen atoms – the hydrogen to oxygen atom ratio is 2:1. This is true for all carbohydrates. The presence of the atoms carbon, hydrogen and oxygen in a carbohydrate can be shown during an experiment in which the hydrogen and oxygen atoms of glucose are removed by adding concentrated sulfuric acid.

During this reaction we can see steam (water) being produced and charcoal (black carbon) is formed.

Starch is a very complex carbohydrate. It is made by joining many, many glucose molecules together – typically around 300–1000. Plants do this to store the glucose made during photosynthesis. Animals obtain starch and glucose by eating vegetables and fruits.

The energy stored in carbohydrates is demonstrated in the Screaming Jelly Baby experiment. A Jelly Baby sweet is placed into molten potassium chlorate. As it reacts, sound, light and heat energy are produced.

carbohydrate

carbon   hydrogen   oxygen

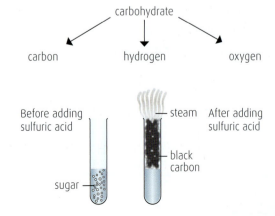

Before adding sulfuric acid

steam   After adding sulfuric acid

black carbon

sugar

During digestion, the starch in the food we eat is broken down into glucose. Glucose molecules are small enough to pass through the lining of the gut into the bloodstream where they are transported to cells. Starch is too big a molecule to pass through the gut lining and so it is broken down into smaller glucose molecules. Acid and enzymes present in the gut speed up this reaction.

In cells, the glucose is broken down by enzymes into carbon dioxide and water. This process is known as **respiration** and releases energy into the cell.

$$\text{glucose + oxygen} \rightarrow \text{carbon dioxide + water}$$

 TASK

Carbon dioxide needs to be removed from the cell after respiration. Try to find out how carbon dioxide is removed.

TESTING FOR STARCH AND GLUCOSE

Benedict's test

Some simple sugars, like glucose, react with blue Benedict's solution to produce a brick red (or orangey) cloudy solution.

There is one noteable exception – the simple sugar sucrose (normal table sugar) will **not** react with Benedict's solution and a mixture of Benedict's solution and sucrose remains blue after heating.

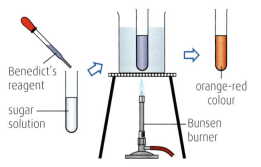

Benedict's reagent

sugar solution

orange-red colour

Bunsen burner

Iodine test

Iodine can be used to detect the presence of starch. When the orangey brown iodine solution is added to starch a blue–black colour appears. In the picture on the right, iodine has been used to show the presence of starch in a seed.

 THINGS TO DO AND THINK ABOUT

1. What is photosynthesis?
2. Write a word equation for photosynthesis.
3. Name three substances that plants make from glucose.
4. Name the three elements present in a carbohydrate.
5. Write the molecular formula for glucose.
6. Describe an experiment to show that sugar contains carbon, hydrogen and oxygen.
7. What is the ratio of hydrogen to oxygen atoms in a carbohydrate?
8. What type of substance is starch?
9. Why do plants make starch?
10. Name the process during which glucose is broken down to release energy in cells.
11. Describe the test for glucose. You should remember to include what you would expect to see in your description.
12. Describe the test for starch. You should remember to include what you would expect to see in your description.

Test-tube containing iodine solution.

Test-tube containing iodine solution and starch.

ALCOHOL

Alcohol is another important consumer product that can be produced from plants. As well as being used in alcoholic drinks, alcohol is used as a disinfectant, a fuel and has many other uses.

FERMENTATION

Fermentation is the name of the reaction that converts glucose in plants into the alcohol, ethanol. For this reaction to take place, yeast (a fungus) is needed.

glucose → ethanol + carbon dioxide

Enzymes present in yeast catalyse the reaction. The reaction takes place in the absence of oxygen in warm conditions (around 25–30°C).

Fermentation experiments can be carried out in the laboratory similar to the one shown in the diagram.

After a few days of fermentation, the lime water in the second boiling tube turns cloudy, showing that carbon dioxide has been produced. The alcohol present in the reaction mixture is relatively dilute. In fact, only alcohol concentrations of about 14% can be produced by fermentation as the alcohol produced kills the yeast.

Alcoholic drinks

There are many different types of alcoholic drink and there are also many different plants that can be fermented. The table below lists a few.

70% alcohol hand sanitizer

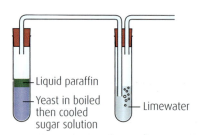

An experiment to show that carbon dioxide is produced during fermentation. The sugar solution is first boiled before being used to remove oxygen from the solution. The paraffin on the top of the sugar solution stops oxygen in the air from dissolving in the solution.

Plant used in fermentation	Alcoholic drink
Grapes	Wine
Barley and hops	Beer
Potatoes	Vodka
Barley	Whisky

TASK

The alcohol in gin can be made by fermenting molasses (dark treacle) or grain. Try to find the name of the plant that is used to give gin its distinctive flavour.

DISTILLATION

Some alcoholic drinks, such as vodka, whisky and gin, have alcohol concentrations greater than 14% and so cannot be produced only from fermentation. These types of drinks are generally called 'spirits'. To achieve higher alcohol concentrations, a process called **distillation** is used after fermentation. Distillation separates the ethanol from the water in the fermentation mixture. In the laboratory, distillation is carried out as shown in the diagram on page 36.

Ethanol and water have different boiling temperatures. The boiling point of ethanol is 78°C and that of water is 100°C. Heat is supplied to the mixture until the ethanol begins to boil and turns into a gas (**evaporates**). As the ethanol vapour passes the thermometer, the temperature measured will correspond to the temperature of the ethanol vapour (around 78°C). The gas then enters the **condenser** and **condenses** back into a liquid, drips down the condenser and collects in the conical flask. This liquid is called the **distiliate**. In the experiment on page 36, the distillate will be ethanol.

LABELS AND UNITS

As we have seen, alcoholic drinks vary widely in the concentration of alcohol they contain. The concentration of a drink is stated on the label.

This label states that the alcohol in this drink makes up 12% of the volume of the bottle. In other words, there are 12 cm³ of ethanol in every 100 cm³ of the drink.

Whisky may contain around 40% alcohol, whereas beer contains around 6%. These different concentrations of alcohol make it harder to know how much pure alcohol is in a given volume of a drink. To make this easier, the alcohol concentration can also be expressed in units. In the UK, one unit is 10 cm³ of pure alcohol. The label above shows that this particular drink contains 9 units of alcohol.

The government have produced safe drink limits for maintaining good health.

A woman should have no more than two 125 cm³ glasses of the alcoholic drink shown in the label above.

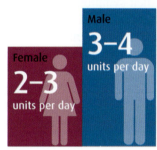
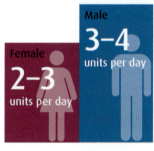

DON'T FORGET

Fermentation only produces alcohol concentrations of around 14%. Above this concentration the yeast is killed.

DON'T FORGET

Alcohol is a drug and has harmful effects on both your body and your brain.

The picture above shows a 'still' used to produce whisky. These tend to be made from copper, unlike laboratory stills which are made from glass.

THINGS TO DO AND THINK ABOUT

1. Write the word equation for fermentation.
2. What else needs to be added to glucose for fermentation to take place?
3. Describe a test that can be carried out to show that the gas produced during fermentation is carbon dioxide.
4. Draw a table with the headings **Plant used in fermentation** and **Alcoholic drink**. Use the information in this section to fill in the table showing four different alcoholic drinks.
5. Why can fermentation only produce alcohol concentrations of around 14%?
6. What process allows alcohol concentrations of greater than 14% to be produced?
7. Explain how ethanol and water are separated during distillation.
8. What is meant by one **unit** of alcohol?

PLANTS TO PRODUCTS

We have already learned some of the many uses of plants. As well as being an essential requirement in our diet, plant products are useful as fuels and can be used to make many other consumer products such as alcohol and paper. In this section we will look more closely at a few plants and learn about their specific uses.

MEDICINES: ASPIRIN

In the previous section we learned of the many different plants that can be used to produce alcohol. Alcohol is classified as a drug due to its effect on the body. Medicines are also drugs, but the term medicine is used to describe a drug that is used to treat diseases or ailments. Approximately 30% of all medicines are derived from plants and around 25% of all prescription medicines contain materials isolated from plants. Many other medicines have been made by chemists by copying a plant chemical and modifying it to make it more suited to its purpose.

An example is aspirin. It is a non-prescription medicine used to treat pain and to help reduce fever.

Name of plant

Willow

It was discovered in 1763 that the bark of willow trees (see photo on left) helped to reduce fever. Later a Scottish doctor discovered that the bark of willow trees also helped to reduce the symptoms of rheumatism (the name of any disease that causes inflammation in joints, muscles or soft tissue).

Chemists discovered that willow bark and the flowers of the meadowsweet plant, spiraea (shown on the right), contain the same compound.

Where it is found

Willow originates from China, but can now be found in Europe, North America and Asia.

Meadowsweet, or spiraea, is common throughout Europe and can also be found in the eastern USA and Canada.

Active ingredients

It was found that both willow bark and spiraea contained a chemical called salicin. It was then discovered that salicin breaks down into salicylic acid (2-hydroxybenzoic acid) in the body. The structure of salicylic acid is shown below.

Role of chemists

Meadowsweet spiraea

As well as isolating the compound salicin from willow bark, chemists have been involved in modifying salicylic acid to make it more suitable for use as a medicine. When salicylic acid was first given to patients it was found that, although their symptoms were reduced, they suffered severe irritation to the mouth, gullet and stomach. The new compound, sodium salicylate, was equally effective at reducing pain and fever but tasted terrible and caused patients to vomit. Chemists continued to work on making a better medicine and in the 1890s a compound was made that worked well as a medicine, had an improved taste and caused less irritation to the mouth, gullet and stomach. This new compound was acetylsalicylic acid and was given the name aspirin ('a' for acetyl and 'spiri' for spiraea, the plant from which the first active ingredient was isolated).

Uses and applications

Aspirin is mainly used to help reduce fever, inflammation and pain, although more recently it has also been linked to reducing the occurrence of heart disease and strokes.

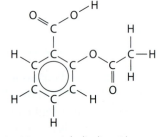

Salicylic acid – the active ingredient in willow bark and spirea

Aspirin – acetylsalicylic acid

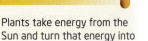

DON'T FORGET

Plants take energy from the Sun and turn that energy into substances that can be used to make useful products.

Benefits to everyday life

Aspirin is a regularly used, relatively safe non-prescription medicine used for a wide range of conditions. In the past, many people, especially children, would have died as a result of a fever. Medicines such as aspirin have meant that, thankfully, death resulting from a fever is now relatively rare in developed countries like Scotland.

TASK

Can you find out about another medicine that is plant based? Try to find the name of the plant, where the plant is found, what the active ingredient is, the role of chemists, its uses and its benefits to everyday life.

SOAPS AND COSMETICS

There are many different plant based compounds in soaps and cosmetics. Often plant extracts are used as perfumes and colourings. An example is an essential oil called limonene, which is extracted from the peel of oranges and lemons and is responsible for the citrus smell of many soaps and shampoos.

DYES AND FOOD COLOURINGS

Have you noticed that many fruit and vegetables produce a highly coloured juice? Examples include blackberries, blueberries and raspberries. The juice from these fruits can be used to dye cloth – a process that has been used for hundreds of years to produce coloured tartan. Dyes can also be made from lichen, tree bark, plant roots, or from the leaves of plants and trees. Dyes from plants are also often used as food colourings.

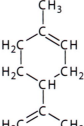

Structure of the chemical, limonene, which is used to give shampoos a citrus smell.

Plant	Colour of dye
brambles	purple
blueberries	purple
raspberries	red
gorse flowers	yellow
nettle leaves	green

TASK

Try to find a method to dye a piece of cloth using natural plant dyes. What name is given to the chemical that is used to 'fix' the colour permanently to the cloth?

THINGS TO DO AND THINK ABOUT

1. Copy and complete the table for two different plant based medicines.

	Medicine 1	Medicine 2
Name of medicine		
Name of plant		
Where is the plant found?		
Active ingredient		
Role of chemists		
Uses and applications		
Benefits to everyday life		

2. Glycerin is a plant product that is found in many shampoos and cosmetics. Can you find out what plants it comes from and its use in shampoos and cosmetics?

3. Many dyes and food colourings are plant based. Name the plants which could be used to make a tartan that had the colours green, yellow, red and purple.

KEY AREA QUESTIONS

FUELS

1. Coal, oil and natural gas are examples of fossil fuels.

    (a) State what is meant by the term **fossil** when used to describe these fuels.

    (b) State one disadvantage of using fossil fuels.

    (c) Burning of fossil fuels is an example of an exothermic reaction. State what is meant by the term **exothermic**.

    (d) The table below shows the energy released by some fuels.

Fuel	Average Energy (kJ/kg)
Wood	15900
Peat	17150
Petrol	48000
Diesel	44800

    (i) Name the fuel that produces most energy per kg.

    (ii) A tank of petrol holds 60 litres and 1 litre of petrol has a mass of 0.75 kg.
    Calculate the total mass of petrol in a 60 litre tank of petrol.

    (iii) Calculate the total energy that would be given out if all the petrol in the tank was burned.

2. Fossil fuels consist mainly of hydrocarbon compounds.

    (a) Name the elements present in a hydrocarbon.

    (b) Write a word equation for the complete combustion of a hydrocarbon.

    (c) Describe a test that could be used to show that carbon dioxide is formed when a hydrocarbon fuel burns.

    (d) Name the poisonous gas produced when a hydrocarbon burns in an insufficient supply of oxygen.

    (e) Fossil fuels are not carbon neutral. Name a fuel that is said to be carbon neutral.

3. Biofuels are made from biomass.
    There are many different types of biofuel. Name and state a use for one biofuel.

HYDROCARBONS

1. Hydrocarbon compounds can be obtained by fractional distillation of crude oil.

    (a) Name the property that is used to separate hydrocarbons by distillation.

    (b) Name the fraction from crude oil that has the lowest boiling point.

    (c) State a use for the residue fraction from crude oil.

    (d) Name the fraction from crude oil that is made up of hydrocarbon molecules ranging between 5 and 9 carbon atoms.

    (e) There is a high demand for the fraction containing hydrocarbon molecules of 5–9 carbon atoms.
    Name the process used to break larger hydrocarbon alkane molecules into smaller, more useful molecules.

2. The alkanes are a family of hydrocarbon compounds.

    (a) Name the alkane that contains four carbon atoms per molecule.

    (b) Write the molecular formula for propane.

    The boiling points of some alkanes are given in the table

Alkane	Carbon atoms per molecule	Boiling point (°C)
Pentane	5	36
Hexane	6	69
Heptane	7	98
Octane	8	126
Nonane	9	?

    (c) Predict the boiling point of **nonane**.

    (d) Name the alkane shown:

    (e) The alkenes are another hydrocarbon family.
    Describe the chemical test, including the result, for an alkene.

EVERYDAY CONSUMER PRODUCTS

1. Starch is a complex carbohydrate made by plants.

    (a) Name the simple carbohydrate that starch is made from.

    (b) State the reason for plants making starch.

    (c) Describe a chemical test that would confirm the presence of starch in a substance.

    (d) Benedict's solution can be used to test for the presence of glucose.
    Describe the colour change that would occur if glucose was tested with Benedict's solution.

    (e) Starch is broken down in the body. State the name of the process that breaks down starch in the body.

    (f) State the reason for the body breaking down starch.

2. Simple carbohydrates, like glucose, can be fermented to produce ethanol.

    (a) Name the other product that is formed when glucose is fermented.

    (b) Name a substance that must be added to glucose for fermentation to occur.

    (c) Only alcohol concentrations of around 14% can be made by fermentation. State the name of the process that is used to produce alcohol concentrations greater than 14%.

    (d) Other than the percentage of alcohol content, state another method used to measure the alcohol content of drinks.

    (e) An alcohol concentration of 1% means that there is 1 cm^3 of pure alcohol in 100 cm^3 of the alcoholic drink. 7% means that there are 7 cm^3 of pure alcohol per 100 cm^3 of alcoholic drink. Calculate the volume of pure alcohol that would be in a 750 cm^3 bottle of wine that had an alcohol concentration of 11%.

PLANTS TO PRODUCTS

1. Many useful everyday products are derived from plants. Aspirin is a medicine that chemists made by copying an active ingredient, salicin, found in plants.

    (a) Name a plant that contains the active ingredient, salicin.

    (b) Salicin is no longer used as a medicine. Give an advantage of using aspirin over salicin.

    (c) Other than medicines, name another useful product derived from plants.

CHEMISTRY IN SOCIETY

METALS AND ALLOYS

METALS

There are 118 chemical elements in the Periodic Table. Of these 118 elements, 94 are metals, which can be found on the left-hand side of the dividing step drawn on most Periodic Tables.

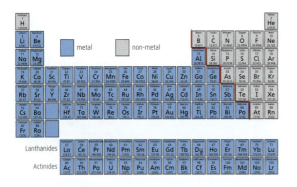

Metallic elements include copper, iron, magnesium, silver and gold.

PROPERTIES OF METALS

Metals are very useful materials as they have a wide variety of uses, which are linked to their properties.

Uses of metals

Copper metal is used as the material for making electrical cables because it is a good conductor of electricity and also because it is **ductile** and can be stretched into wire.

Aluminium metal is used to make bicycle frames and parts of the body of an aeroplane. This is because it is strong, light and does not **corrode**.

Silver, gold and platinum metals are used to make jewellery and coins. This is because they are **malleable**, shiny and very unreactive.

DON'T FORGET

In the section on 'Chemical changes and structure' you learned that the elements can be classified as metals or non-metals.

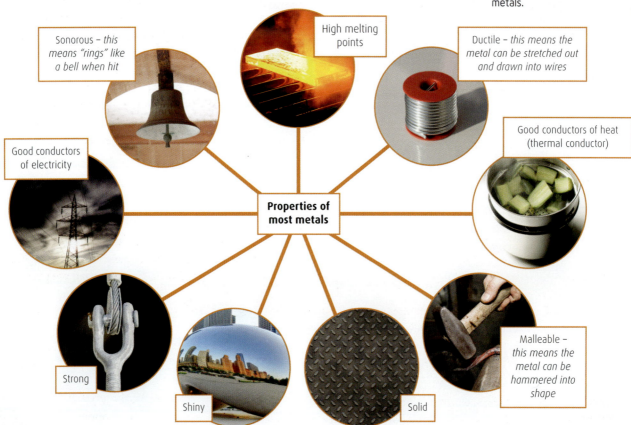

Sonorous – *this means "rings" like a bell when hit*

High melting points

Ductile – *this means the metal can be stretched out and drawn into wires*

Good conductors of electricity

Good conductors of heat (thermal conductor)

Properties of most metals

Strong

Shiny

Solid

Malleable – *this means the metal can be hammered into shape*

Iron metal is used to make many different objects and building structures (fences, gates and bridges) because of its properties. Unfortunately, when iron metal reacts with air and water it corrodes to form rust on its surface, which makes it weaker and brittle.

ALLOYS

The properties of metals can be changed and improved by forming alloys. An **alloy** is a mixture of two or more elements, at least one of which is a metal.

Alloying metals produces new materials that exhibit more desirable properties. The metals that are selected for use in the alloy depend on the properties that are required in the resulting alloy.

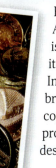

British coins are made of alloys. A £1 coin, though gold coloured, is not made of gold because gold itself is too soft and expensive. Instead it is made from a nickel–brass alloy, which contains 70% copper, 24.5% zinc and 5.5% nickel, producing a harder coin with the desired colour.

Copper coins, such as the 1p and 2p pieces, used to be made from bronze alloys which contained copper, tin and zinc. In 1992, as a result of the increasing price of copper, the Royal Mint replaced this bronze alloy with the alloy steel, which contains iron. Steel is plated with a thin layer of copper to produce the 'copper' finish. This use of steel explains why coins made after 1992 are magnetic.

Eighteen-carat gold used for jewellery is not made from pure gold as pure gold is too soft, but is an alloy of gold mixed with copper, which produces a harder metal.

Steel, which is used for railway tracks and car body parts, is an alloy made from iron mixed with the non-metal element carbon. This produces a much stronger material, which is also less brittle than iron alone. It does, however, still corrode to form rust.

On the other hand, stainless steel, which is used to produce cutlery and kitchen sinks, is an alloy of iron, carbon and chromium. It is not only strong, but also produces a more shiny material that is more resistant to corrosion than traditional steel.

Shape memory alloys

In the 1960s a new type of alloy was developed called a shape memory alloy. An alloy of the metals nickel and titanium, it keeps the memory of its original shape. The alloy can be bent or stretched to change its shape, but will return to its original shape on heating.

Originally developed for the military and NASA, these shape memory alloys now have many everyday practical applications. They form springs in greenhouse ventilation systems so the roof panels will open and shut automatically when the temperature changes. They are also used in dental braces, as electrical components that trigger fire alarm systems and are also used to make the frames of spectacles so that even if the frames become misshapen, they can be returned to their original shape on heating.

THINGS TO DO AND THINK ABOUT

1. Use a copy of the Periodic Table to identify the name of the metal from the following chemical symbols:
   (a) Au   (b) Cu   (c) Mg   (d) Ni   (e) K
2. Use a Periodic Table to identify the chemical symbol for each of the following metals:
   (a) Zinc   (d) Calcium
   (b) Silver   (e) Iron
   (c) Aluminium
3. Iron is the most widely used metal.
   (a) Write a list of as many uses of iron that you can think of.

(b) Now put an asterisk * beside any of the uses you have identified where steel, which is an alloy of iron and carbon, might be used rather than the pure iron metal.

4. The properties of different metals are linked to their uses.
   Give a use for each of the following metals and identify a property for each metal that explains this use.
   (a) Iron   (d) Mercury
   (b) Silver   (e) Copper
   (c) Aluminium

REACTIONS OF METALS

Some metals are very reactive and some metals are very unreactive.

By observing how metals react in oxygen, water and dilute acid, chemists have been able to produce an 'order of reactivity' for the metals called the reactivity series.

REACTION WITH OXYGEN

Magnesium burns in oxygen with an extremely bright white flame, producing a white powdery solid of magnesium oxide.

When metals are heated in oxygen they can react and combine with the oxygen to form a metal oxide:

> metal + oxygen → metal oxide

For example, magnesium burns in oxygen to produce magnesium oxide.

The word equation for the reaction is:

> sodium + oxygen → sodium oxide

Sodium oxide is an ionic compound.

Sodium burning in oxygen

REACTION WITH WATER

When metals react with water they produce a metal hydroxide and hydrogen gas.

For example, lithium reacts vigorously with water, producing the flammable gas hydrogen and the alkaline compound called lithium hydroxide.

> lithium + water → lithium hydroxide + hydrogen

Magnesium metal, however, reacts very slowly in water.

Metals such as silver and gold are unreactive and do not react in water at all.

This is one of the chemical properties of silver and gold that makes them suitable materials for jewellery.

Magnesium burning in oxygen

Iron burning in oxygen

REACTION WITH DILUTE ACID

The alkali metals are so reactive that it would be too dangerous to place them into dilute acid.

Less reactive metals like zinc react with dilute hydrochloric acid to produce a metal chloride and hydrogen gas.

> zinc + hydrochloric acid → zinc chloride + hydrogen

burning splint

magnesium — zinc — copper

dilute HCl —

The burning splint is used to test for hydrogen gas which is present above the test tubes containing magnesium and zinc.

58

REACTIVITY SERIES OF METALS

The **reactivity series** of metals places metals in order of their reactivity starting with the most reactive at the top and the least reactive at the bottom.

The table below shows the order of reactivity and provides a summary of the reactions of metals.

Metal	Reaction with oxygen	Reaction with water	Reaction with dilute acid	
potassium sodium lithium calcium magnesium aluminium zinc iron tin lead copper silver gold	Vigorous reaction *metal + oxygen* → *metal oxide* Very slow reaction No reaction with oxygen	Vigorous reaction *metal + water* → *metal hydroxide + hydrogen* Very slow reaction No reaction with water	Violent reaction *metal + dilute acid* → *salt + hydrogen* Very slow reaction No reaction with dilute acid	Most reactive ↑ ↓ Least reactive

DON'T FORGET

Ionic compounds form when metal atoms join to non-metal atoms by transferring electron(s) from the metal to the non-metal. The resulting charged particles are called ions and an ionic bond is formed by the attraction of the oppositely charged ions.

DON'T FORGET

You should be able to place metals correctly in the reactivity series and make predictions about the reactions of metals based on their position in the reactivity series.

THINGS TO DO AND THINK ABOUT

1. Use the reactivity series to explain why copper metal is used to make water pipes, but sodium metal would not make a suitable material for water pipes.

2. The chemicals found in the fireworks called sparklers burn and react with oxygen from the air.

   One of the elements found in sparklers is the metal iron.

   Write a word equation for the chemical reaction that takes place when iron metal reacts with oxygen.

3. The reactivity series starts with potassium, the most reactive metal in the series, and finishes with gold, the least reactive metal in the series.

   Use a copy of the Periodic Table to rewrite this reactivity series using the chemical symbols for each of the metals rather than the chemical names.

4. X, Y and Z are three metals.

   Metal X reacts with dilute acid, but does not react in water.

   Metal Y does not react with dilute acid or with water.

   Metal Z reacts with both dilute acid and water.

   (a) Use this information to place metals X, Y and Z in order of reactivity, starting with the most reactive.

   (b) Using the information given in the reactivity series at the start of this section, suggest a name for each of the metals X, Y and Z.

Copper pipes

Sparklers contain iron

59

EXTRACTION OF METALS

METAL ORES

Metals have a variety of uses depending on their physical and chemical properties.

Most metals are not found in nature as the element, but instead exist as metallic compounds called ores. An ore is a naturally occurring metal compound found in rocks or under the ground.

Examples of ores are shown in the table below:

Name of ore	Metal found in the ore
Haematite	Iron
Bauxite	Aluminium
Iron pyrite (fool's gold)	Iron
Malachite	Copper

Silver metal trapped in rock

EXTRACTION OF METALS

Unreactive metals found at the bottom of the reactivity series such as gold and silver are found naturally in the ground as the element.

Other more reactive metals must first be extracted from their ores. Extraction involves separating the metal from the other elements contained in the ore.

The method used to extract a metal from its ore depends on the position of the metal in the reactivity series. In general, the less reactive a metal is (the lower in the reactivity series the metal is found), the easier it is to extract the metal from its ore.

METHODS OF EXTRACTION

Using heat alone (silver)

Silver is quite unreactive but can be found as sulfides or oxides. Heat energy is sufficient to break up the compound separating the silver from the oxygen allowing the metal to be extracted.

Unreactive metals

Gold is very unreactive and is found in rock as the native metal. For this reason, it does not need to be separated from the rock using chemical methods.

Heating with carbon or carbon monoxide (copper, lead, tin, iron and zinc)

Heat alone is not sufficient to provide the energy required to extract the metals found in the middle of the reactivity series from their ores. Instead the ores must be heated in the presence of carbon or carbon monoxide.

The carbon or carbon monoxide removes the oxygen from the ore and is **oxidised** to form carbon dioxide.

DON'T FORGET

Unreactive metals are found at the bottom of the reactivity series.

DON'T FORGET

An element is one kind of atom. An element cannot be broken down into anything simpler.

Haematite

Bauxite

Iron pyrite (fool's gold)

Malachite

DON'T FORGET

Reactivity series – potassium, sodium, lithium, calcium, magnesium, aluminium, zinc, iron, tin, lead, copper, silver and gold.

Gold mine in Cononish, near Tyndrum, in the Loch Lomond National Park

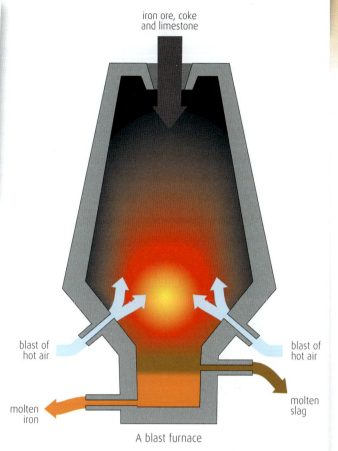

A blast furnace

Extracting iron metal

Iron metal is extracted from haematite ore in a blast furnace.

The word equation for the main reaction that takes place in the blast furnace is:

iron oxide + carbon monoxide → iron + carbon dioxide

Electrolysis of molten ores (aluminium, magnesium, calcium, lithium, sodium and potassium)

The more reactive metals are very strongly bonded to the other elements in their ores.

More energy is required to separate the ores and this is provided by **electrolysis.** Electrolysis is a process that separates a compound into its elements using electrical energy (electricity).

Aluminium metal is extracted from its molten ore using electrolysis.

Electrolysis of aluminium

Iron ore, coke (a form of carbon) and limestone (calcium carbonate) are added to the blast furnace.

It is called a blast furnace because of the blasts of hot air which are used to increase the temperatures inside the furnace and to supply oxygen.

The carbon (coke) reacts with this oxygen to form, in the first instance, carbon dioxide, which then further reacts with more of the carbon to form carbon monoxide.

The carbon monoxide reacts with the iron oxide (ore) reducing it to iron metal and is itself oxidised to carbon dioxide gas.

The high temperatures in the furnace produce the extracted iron in the form of a **molten** liquid.

THINGS TO DO AND THINK ABOUT

1. Produce a summary table to match the position of a metal in the reactivity series to the method used to extract it from its ore.

2. Platinum is a precious metal used to make jewellery. It can be extracted from its ore using heat alone.

   What does this suggest about the reactivity of platinum?

3. Copy and complete the following table:

Name of ore	Chemical name	Chemical formula
Haematite		Fe_2O_3
Bauxite	Aluminium oxide	
Iron pyrite (fool's gold)		FeS_2
Malachite		$CuCO_3$

A nugget of platinum

CORROSION OF METALS

CORROSION

Corrosion is an everyday chemical reaction that takes place on the surface of a metal. The metal reacts with water and oxygen from the air to form a metal compound.

Corrosion is an example of a type of chemical reaction called oxidation. An oxidation reaction takes place when a metal reacts with oxygen from the air to form a metal oxide.

Metals corrode at different rates depending on their position in the reactivity series.

The corrosion of a metal is not a useful chemical reaction because the corroded metal is weaker and more brittle than the pure metal.

This picture shows potassium metal, which is very reactive and is stored under oil to prevent it reacting with air and water. As soon as it is exposed to air it loses its shiny appearance and turns greyish black. This is because the surface of the potassium is corroding and the surface of the metal reacts to become potassium oxide.

CORROSION OF IRON

Iron metal reacts with both water and oxygen in the atmosphere to produce the compound iron oxide.

Rusting is a term that can only be applied to the corrosion of the metal iron because rust is another name for iron oxide.

Iron oxide (rust) is a brittle compound that weakens the strength of the original piece of iron or steel.

In tube 1 the nail did not rust because it was only exposed to water. In tube 2 the nail did not rust because it was only exposed to air (oxygen). In tube 3 the nail rusted because it was exposed to both air and water.

For rusting to take place, both air (containing oxygen) and water must be present, as shown in the following experiment.

Ferroxyl indicator is a chemical indicator that can show us when rusting is taking place.

The indicator is pale yellow in colour, but changes colour to bright blue if rusting of iron occurs.

In 2015, the statue of liberty in New York was 129 years old. Made from the less reactive metal copper, it was not always this colour. Over many years the copper metal has been exposed to the air and water. The copper statue corroded, forming a green coloured covering known as verdigris, which we associate with the statue today.

PREVENTING CORROSION

Chemists have developed many different methods to prevent the corrosion of metals.

Physical barriers

Remember that for a metal to corrode it must be exposed to both air and water.

One way to prevent corrosion of a metal is to put a physical barrier around the metal that keeps out both air and water.

Physical methods used include:

- Painting the metal

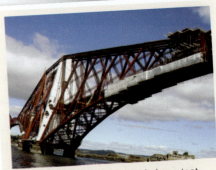

The Forth Rail Bridge has needed constant painting to protect the steel structure from corrosion since its completion in 1890. In 2011 this finally came to an end with the completion of a £130 million pound, 10-year project to cover the bridge in a special anti-corrosion coating.

- Coating the metal in a layer of plastic
- Covering the metal in oil or grease – this method is used on moving parts like a bicycle chain, for example.
- Galvanising the metal (covering the metal with a layer of zinc)
- Electroplating (less reactive metals are plated onto the surface of the metal being protected using electricity). One example of this is **tin-plating**.

CHEMICAL PROTECTION

When a metal corrodes it changes from a chemical element to a compound. The higher up a metal is in the electrochemical series, the more easily it corrodes.

Sacrificial protection

This method works by taking a metal higher up the reactivity series and connecting it to a less reactive metal to protect the less reactive metal from corrosion.

The more reactive metal will corrode first and is **sacrificed** for the less reactive metal.

This is shown in the following experiment shown in the diagram on the right:

In petri dish 1, the blue colour that forms around the nail wrapped in copper indicates rusting is taking place. Iron is higher up in the electrochemical (and reactivity) series than copper and therefore corrodes, sacrificing itself to protect the copper.

In petri dish 2, corrosion also takes place, indicated by the ferroxyl indicator, which has turned blue, and this is because the nail is unprotected from rusting.

The nail in petri dish 3 has not corroded, indicated by the fact that the ferroxyl indicator has not turned blue. Magnesium is higher up in the electrochemical (and reactivity) series than iron and this means that the magnesium will corrode and be sacrificed to protect the less reactive iron nail.

This clever chemistry can be used to prevent corrosion in much larger metal structures such as the hulls of ships.

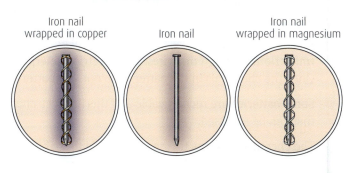

Iron nail wrapped in copper | Iron nail | Iron nail wrapped in magnesium

The hull of a giant ship.
Blocks of zinc metal have been bolted onto the hull of the iron/steel ship to prevent the hull from rusting.
As zinc is higher up in the electrochemical series than iron, the zinc corrodes, sacrificing itself to protect the iron/steel from rusting.
The zinc blocks, which corrode, must be replaced periodically to maintain the sacrificial protection they provide.

THINGS TO DO AND THINK ABOUT

1. What is meant by the term corrosion?
2. Oil rigs like this one in the North Sea are giant structures made from steel. Blocks of metal can be bolted onto the steel to prevent corrosion.
   (a) Suggest two ways in which physical barriers could be used to protect an oil rig.
   (b) Suggest the name of a metal that could be used to protect the steel structure and explain the reason for your choice.
   (c) Suggest the name of a metal that would not be suitable to protect the steel from corrosion and explain why it is not suitable.

3. Physical barriers and chemical protection are both used to protect metals from corrosion.

   Give some everyday examples of either type of protection in use and identify the type of protection using (P) for physical and (C) for chemical.

   Here are two examples to get you started: painting iron gates and railings (P); magnesium metal bolted onto iron pipes (C).

MAKING ELECTRICITY

BATTERIES

The batteries that are used to power toys, torches, wireless keyboards and remote controls all produce electricity, which comes from the chemical reactions taking place inside the battery.

Chemical energy is converted into electrical energy which flows through the wires. This electrical energy powers the electrical device. Once the chemicals in the battery run out, the battery stops working.

Some batteries are rechargeable and this involves chemical reactions that can be reversed.

The chemical changes taking place inside these batteries are reversed when the battery is recharged and the battery can therefore be used over and over again.

ELECTROCHEMISTRY

Electrochemistry involves chemical reactions that take place between different metals to produce a flow of electrons, or electricity.

Electricity can be made by connecting two different metals together in an electrolyte solution to form a simple cell.

The diagrams on the right show a simple electrochemical cell. The chemical reactions taking place are similar to those taking place inside a battery.

Electricity is produced when those two metals are placed in the sodium chloride solution and the electric current flows from the zinc rod to the copper rod through the wires.

A battery is made up of one or more electrochemical cells connected together.

When different pairs of metals are connected together by an electrolyte an electric current is produced because electrons can flow from one metal to the other.

An electrolyte is an ionic compound that can conduct electricity when it is molten (melted) or when dissolved in water to form an aqueous solution for – for example, sodium chloride (common salt) solution.

A fruit battery is a simple electrochemical cell. The juice inside the fruit acts as an electrolyte. Two different metals, for example zinc and copper, are used as electrodes. When the metal electrodes are connected in the cell a chemical reaction takes place.

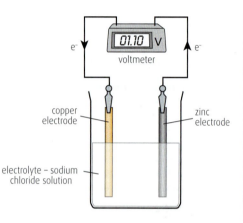

DON'T FORGET

Electricity is produced when there is a flow of electrons from one metal to another through connecting wires.

EXAMPLE

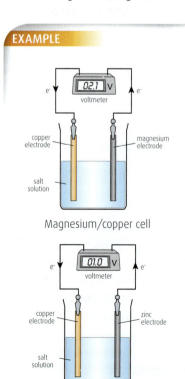

Magnesium/copper cell

Zinc/copper cell

A fruit battery

Zinc is more reactive than copper and gives up some of its electrons, which travel through the wires to the copper electrode producing electricity.

Different pairs of metals produce different voltages.

This difference is used to place metals in their correct order within the **electrochemical series.**

The electrochemical series is a list of both metals and non-metals and places the more reactive metals at the top and the least reactive metals at the bottom.

In an electrochemical cell, electrons travel from a metal higher up in the electrochemical series to a metal lower down in the electrochemical series.

Magnesium is higher up the electrochemical series than copper and therefore electrons will travel through the wires from the magnesium metal to the copper metal.

The further apart the metals are from each other in the electrochemical series the greater the voltage produced by the electrochemical cell.

The zinc/copper cell produces a lower voltage than the magnesium/copper cell because magnesium and copper are further apart in the electrochemical series than zinc is from copper.

Displacement reactions

If a metal higher up the electrochemical series is added to an aqueous solution of an ionic compound containing a metal lower down the electrochemical series, then a chemical reaction called **displacement** takes place.

The metal that is higher in the electrochemical series displaces the metal that is lower in the electrochemical series.

Metal
lithium
potassium
calcium
sodium
magnesium
aluminium
zinc
iron
nickel
tin
lead
hydrogen
copper
silver
mercury
gold

EXAMPLE

In the diagram, the iron (nail) is higher up in the electrochemical series than the copper metal found in the copper sulfate solution. The iron displaces (takes the place of) the copper from its compound, copper sulfate, forming an iron sulfate solution and copper metal.

Iron nail — Leave for one week while reaction takes place

blue copper sulfate solution
green iron sulfate solution
Copper metal on iron

Before   **After**

The word equation for this displacement reaction is:

iron (s) + copper sulfate (aq) → iron sulfate (aq) + copper (s)

If the metal used in this experiment had been lower in the electrochemical series than the copper found in the compound, for example silver, then no displacement reaction would have taken place.

silver (s) + copper sulfate (aq) → no reaction

The results of displacement reactions can be used to place metals correctly in the electrochemical series.

THINGS TO DO AND THINK ABOUT

1. Which of the two cells magnesium/copper or copper/silver will produce the greatest voltage? (Hint: You may wish to look at each metal's position in the electrochemical series.)
2. Consider the diagram of the lemon battery below.
   (a) What would happen to the voltage produced by this cell if the copper coin was replaced by (i) an iron nail or (ii) a zinc nail?
   (b) What would happen to the direction of electron flow if the copper coin was replaced by (i) an iron nail or (ii) a zinc nail?

3. Write a word equation for the displacement reaction that takes place between magnesium metal and zinc chloride solution?
4. (a) Suggest the name of a metal that would be involved in a displacement reaction involving zinc sulfate solution and write a word equation for the chemical reaction that takes place.
   (b) State the name of a metal that would not be involved in a displacement reaction involving zinc sulfate solution and explain your answer.

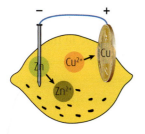

MATERIALS – PLASTICS

WHAT IS A PLASTIC?

Plastics are a group of important **synthetic (artificial)** materials. They are long-chain molecules called **polymers**.

A plastic is made up of small units called **monomers** that when joined together form the long-chain polymer.

Plastics are made through a chemical process called **polymerisation** and have a wide range of properties, hence their many and varied uses.

The name of a polymer can be worked out from the name of the monomer:

Name of monomer	Name of polymer
Ethene	Poly(ethene) also known as polythene
Vinyl chloride	Poly(vinylchloride) also known as PVC
Propene	Poly(propene)
Styrene	Poly(styrene)

THERMOSOFTENING AND THERMOSETTING PLASTICS

Plastics can be divided into two groups depending on how they respond when heated.

If the plastic softens on heating it is a **thermosoftening** (or thermoplastic) plastic. Thermosoftening plastics are flexible and can be remoulded into different shapes using the action of heat.

It is possible to make synthetic fibres from thermosoftening plastics. The plastic is melted and forced through tiny holes, which produces lengths of fibres. The size of the holes can be altered to change the thickness of the fibres produced.

Thermosoftening plastics can be recycled by melting them down and remoulding them into new products.

If the plastic does not soften on heating it is a **thermosetting** plastic. Once moulded into shape the plastics cannot be reshaped. Thermosetting plastics are rigid and tough.

Thermosetting plastics are used for electrical components such as plugs and socket covers.

VARIED PROPERTIES AND USES OF PLASTICS

Plastics have a wide range of properties, which make them suitable for a variety of uses. They are generally lightweight (low density) and waterproof.

They are good heat insulators and also good electrical insulators.

Many plastics are flexible and can be manufactured in many different colours and textured finishes.

Plastics can be engineered to be used in a variety of applications and environments. Here are some examples.

Polystyrene

Expanded poly(styrene) is extremely lightweight and can absorb impact forces, making it suitable as a packaging material to protect electrical appliances. It is also an excellent heat insulator and is used in disposable food packaging and drinking cups.

Poly(styrene) can also be rigid and is used for the cases of CDs and DVDs.

PVC

Poly(vinylchloride) (PVC) is a tough, rigid and waterproof plastic used as the covering for electrical cables and is also used to make the frames of double-glazed windows.

Polythene

Poly(ethene) or polythene is light and strong. It can be formed as a thin film for packaging or to make carrier bags, although it can also be moulded into plastic bottles.

Nylon

Nylon is a strong and flexible material used to make climbing ropes or bristles for toothbrushes, although it can also be drawn into fibres to be woven into clothing.

Even more examples

Newer plastics have been engineered to be water soluble, such as poly(ethenol), and are used in surgical stitches and disposable laundry bags that dissolve in the washing machine.

Kevlar is lightweight and extraordinarily strong, with five times the strength of steel on an equal-weight basis. It is best known for its use in bulletproof and stab-resistant body armour.

Kevlar vest

Magic sand (hydrophobic sand) is coated in water-repellent silicone, not to be confused with the chemical element silicon. Silicones are used as bathroom sealants or in automotive braking systems.

PROBLEMS WITH PLASTICS

Most plastics are very durable and will last for a long time. This is one of their advantages, but is also one of the biggest drawbacks when it comes to their disposal.

Natural materials will decay through the action of microorganisms and are said to be **biodegradable**. Plastics, on the other hand, are synthetic and non-biodegradable so will not decay.

Litter from plastics is an increasing problem as the plastic material remains in the environment for a long time.

The disposal of plastics

We dispose of a lot of plastic by burying it in landfill sites.

Another way some plastics are disposed of is by burning in incinerators. However, this also produces problems as plastics produce poisonous gases when burned. Plastics contain the elements carbon and hydrogen and on burning these elements combine with oxygen in the air to produce new compounds. Most plastics will produce carbon monoxide gas when disposed of in this way. Carbon monoxide is a colourless, odourless gas, which is toxic because it prevents red blood cells from carrying oxygen.

Polyurethane plastics also contain nitrogen and, when they burn, they release the deadly toxic gas hydrogen cyanide (HCN).

Plastics such as poly(vinyl chloride) contain the element chlorine and on burning release a toxic acidic gas called hydrogen chloride.

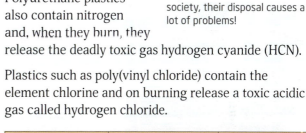

✚ DON'T FORGET

The burning of plastics releases toxic gases. As useful as plastics are to modern society, their disposal causes a lot of problems!

Type of plastic	Toxic gas produced on burning	Molecular formula
General plastics	Carbon monoxide	CO
Poly(vinyl chloride)	Hydrogen chloride	HCl
Polyurethanes	Hydrogen cyanide	HCN

BIODEGRADABLE PLASTICS

Research into biodegradable plastics is widespread as chemists look to solve the problems associated with the disposal of plastic materials in the environment. Biopol was one of the first biodegradable plastics, but there are now various 'green' plastics that use plant material such as cellulose or starch in their structures.

💭 THINGS TO DO AND THINK ABOUT

1. Poly(tetrafluoroethene) is the chemical name for the polymer known as Teflon, which is used as a non-stick coating.
   Name the monomer used to make Teflon.
2. Butene is a monomer that can be polymerised to form a long-chain polymer molecule.
   Name the polymer formed from the monomer butene.
3. A plastic bottle changes shape when filled with hot water.
   Is the bottle made from a thermosoftening or thermosetting polymer?
4. Polymers (plastics) are extremely useful materials.
   (a) Write down the names of three everyday objects that are made from plastic.
   (b) Give a property of a plastic for each of the objects chosen in part (a) that is linked to their use.
5. Plastics are very useful, but also have disadvantages.
   Give three disadvantages of using plastics as a material to make everyday objects.

MATERIALS – CERAMICS AND SMART MATERIALS

CERAMICS

The word ceramic comes from the Greek for potter or pottery.

Ceramics are inorganic and non-metallic materials. They are formed after being exposed to very high temperatures and then cooled, producing solid crystalline materials.

Everyday examples include pottery, ceramic tiles, glass, bricks and cement.

Ceramics have many advantages over other materials such as metals or plastics.

They:

- are hard-wearing
- retain their shape
- are heat resistant.

One disadvantage of ceramics is that, although they are hard and strong, they are also brittle and can break easily on impact.

The useful properties of ceramics have made them vital components for many modern applications.

- Space shuttles - ceramics are used as the material for the tiles that cover the space shuttle as they are very resistant to high temperatures, but also lightweight.
- Hair straighteners - the ceramic plates are hard-wearing, withstand the high temperature and the surface is non-stick.
- Ceramic replacement joints - used in medicine as the material for implants and replacement joints because they are hard-wearing and will not react with anything inside the body.

SMART MATERIALS

Smart materials, include plastics, metal alloys and ceramics, have properties that react to a change in their environment. This means that one of their properties can be changed by an external condition, such as heat, light, pressure or electricity. This change is reversible and can be repeated many times.

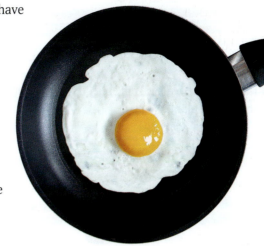

There is a wide range of different smart materials. Each material offers different properties that can be changed.

Examples include:

- plastics that can conduct electricity
- photo chromatic pigments that change colour on exposure to light (used in glasses that change colour depending on light levels)
- shape memory alloys
- thermochromic plastics that change colour on heating (used in baby-feeding spoons that change colour when hot).

THINGS TO DO AND THINK ABOUT

1. Write down everyday examples of ceramics that you might find in the home.

2. Give two properties of bricks that explain their use as building materials.

3. Ceramics are used to coat the inside of frying pans.

   Suggest two properties that explain their use for this purpose.

4. Suggest some other uses of thermochromatic and photochromatic plastics.

FERTILISERS

NATURAL AND SYNTHETIC

Plants for food

All of the food that we eat is produced from either animals or plants.

The world's population is increasing in number and as a result there is an ever-growing demand for food. Farmers have to grow more crops to meet the increasing demand for plants as food.

What do plants need in order to grow?

Plants need water, sunlight, gases from the air, soil and space to grow. There are also three essential elements that plants need: Nitrogen **N,** Phosphorus **P** and Potassium **K.**

Plants absorb nutrients from the soil, which contains the essential elements needed for healthy plant growth. When crops are harvested, these nutrients are removed from the soil and further plant growth can be affected by a lack of nutrients in the soil.

Fertilisers can be added to the soil to replace these lost nutrients to enable farmers to reuse the land immediately to plant more crops.

WHAT IS A FERTILISER?

A fertiliser is a substance that contains the essential elements needed by plants for healthy growth. There are two types of fertiliser.

Natural fertilisers

A natural fertiliser is made from the waste products produced by plants and animals and provides a great source of the element nitrogen. Plant **compost** and animal **manure** are natural fertilisers.

Plant compost

Animal manure

Natural fertilisers are environmentally friendly as they make use of natural waste. However, there are not sufficient quantities of natural fertilisers for them to be the only source of nutrients.

This is where chemistry can play a part in solving everyday problems.

Synthetic (artificial) fertilisers

These are often referred to as **NPK fertilisers** because they are composed of ionic compounds, which contain nitrogen (N), phosphorus (P) and potassium (K). These synthetic fertilisers are salts, which are produced during a neutralisation reaction.

These salts are soluble, which means they will dissolve in damp soil allowing the essential elements to be absorbed through the roots of plants. The name or chemical formula of the salt that makes up the fertiliser often suggests which of the essential elements it provides.

Examples of synthetic fertilisers are shown in the table below.

Name of salt (fertiliser)	Chemical formula	Essential elements (N,P or K)
Ammonium nitrate	NH_4NO_3	N
Ammonium phosphate	$(NH_4)_3PO_4$	N and P
Potassium nitrate	KNO_3	K and N

CALCULATING THE PERCENTAGE COMPOSITION OF A FERTILISER

Farmers need to know how much of each essential element is in a fertiliser so they can select the fertiliser that is most appropriate for their needs.

Calculating the percentage composition of a fertiliser from its packaging

The packaging of a fertiliser indicates the percentage composition of each of the N, P and K elements.

Gro-Green Fertiliser 20-5-10

nitrogen (N) 20% phosphorus (P) 5% potassium (K) 10%

Calculating the percentage composition of essential elements in a fertiliser from its chemical formula

The relative atomic mass (RAM) of the essential element and the number of atoms of this element found in the chemical formula of the fertiliser enables you to calculate its percentage mass composition.

The percentage mass of an element in a compound =

$$\frac{\text{mass of the element in the formula}}{\text{formula mass of the compound}} \times 100$$

EXAMPLE

Calculate the percentage mass of nitrogen in ammonium nitrate. Ammonium nitrate has the chemical formula NH_4NO_3

First, work out the total mass of nitrogen in one formula mass of ammonium nitrate.

There are two atoms of nitrogen in the formula NH_4NO_3

Total mass of nitrogen = 2 x 14 (RAM of nitrogen)

= 28 g

Second, work out the formula mass of ammonium nitrate.

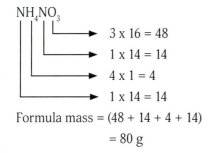

NH_4NO_3
- 3 x 16 = 48
- 1 x 14 = 14
- 4 x 1 = 4
- 1 x 14 = 14

Formula mass = (48 + 14 + 4 + 14)

= 80 g

Third, divide the total mass of the element by the formula mass and multiply by 100.

(total mass of nitrogen / formula mass of ammonium nitrate) x 100

= (28 / 80) x 100

= 35%

Therefore 35% of the formula mass of ammonium nitrate is nitrogen.

Environmental impact

There are problems associated with the use of synthetic fertilisers. As they are soluble in water they can easily wash off the soil and into nearby streams, lakes and rivers.

The fertilisers cause the green algae already in the water to grow at a much faster rate and this algal growth can cover the surface of the water, blocking out sunlight, which then causes other plants in the water to die.

When the algae dies it sinks to the bottom of the water where bacteria break it down. During this decaying process oxygen in the water is used up and this can cause living things, such as fish, in the water to die.

DON'T FORGET

You can use this method to calculate the percentage composition for any given element in any given compound – not just fertilisers.

THINGS TO DO AND THINK ABOUT

1. Name the three essential elements needed for healthy plant growth.
2. Give an example of a natural fertiliser.
3. What is the percentage composition of phosphorus found in a bag of fertiliser labelled as 21-8-11?
4. What is the percentage composition of nitrogen and phosphorus found in fertilisers labelled 29-0-5?
5. Calculate the percentage composition of potassium in the fertiliser potassium nitrate KNO_3?
6. Potassium nitrate salt can be produced from the chemical reaction that takes place between potassium hydroxide (an alkali) and nitric acid. Water is also made during the reaction.
   (a) Name the type of chemical reaction that is taking place.
   (b) Write a word equation for this reaction.

Gro-Green Fertiliser

21-8-11

nitrogen (N) 20%   phosphorus (P) 5%   potassium (K) 10%

NUCLEAR CHEMISTRY

THE ORIGIN OF THE ELEMENTS

The Big Bang theory is the scientific explanation for the creation of our universe.

It is thought that all matter was once all concentrated in a single point and existed as a 'bubble' of hot, dense material.

Then, approximately 13.8 billion years ago, there was a sudden explosion or, as we call it, the Big Bang.

Our universe was created. In a fraction of a second, there was a rapid expansion of matter causing the universe to grow from smaller than a single atom to bigger than a galaxy. It kept on growing and is still expanding today.

After this initial expansion, the first elements were formed in the moments following the Big Bang.

These first elements were the lightest and most abundant elements, hydrogen and helium.

The Sun is our closest star. It is a large ball of hot gases made up of approximately 72% hydrogen, 26% helium and 4% other trace elements. Nuclear fusion reactions take place in the Sun's core converting hydrogen gas into heavier elements. Energy is also released in the form of heat and light.

All the other naturally occurring elements heavier than hydrogen and helium formed later during nuclear fusion reactions taking place inside stars.

Elements are made up of tiny particles called atoms. The extremely high temperatures and pressures found inside a star cause these atoms to join or fuse together and forming new larger and heavier elements.

BACKGROUND RADIATION

We are all exposed daily to a low level form of radiation called background radiation. It cannot be avoided because it is all around us. Background radiation can be natural or artificial.

Natural sources

- Cosmic radiation – this radiation comes from outer space, from the galaxies, stars and black holes. Our Sun also produces background radiation.
- Food and drink – some naturally occurring elements contain radioactive atoms and these can be found in very small quantities in some food and drink, for example in shellfish and in Brazil nuts.
- Radon gas – rocks and soil can emit the radioactive gas called radon.
- Living things – plants can take up radioactive substances through their roots from the soil. This allows the radiation to enter the food chain where it will be ingested by animals, including ourselves.

Artificial sources

Human activity has produced artificial background radiation through the creation and use of artificial sources of radiation such as:

- radioactive waste from nuclear power stations
- nuclear weapons testing and nuclear accidents
- coal-fired and nuclear power stations both emit radioactive particles into the atmosphere
- the use of X-rays in medicine, which is the largest contributor to artificial radiation.

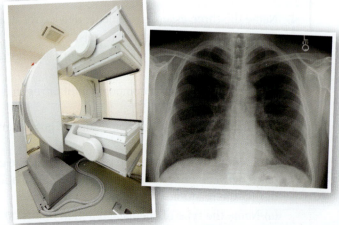

Henri Becquerel

Pierre and Marie Curie

A Geiger counter

WHAT IS RADIOACTIVITY?

In 1896 the French scientist Henri Becquerel discovered the invisible phenomenon we know today as radiation. His experiments and those of fellow scientists Pierre and Marie Curie led to the discovery of radioactivity.

In 1903 Becquerel and the Curies were awarded a joint Nobel Prize for Physics in recognition of the importance of their work.

Radioactivity is the result of the emission of atomic particles and energy from unstable atoms.

These unstable atoms decay to form new atoms and this will continue until the new atoms formed are themselves stable.

This **radioactive decay** is completely spontaneous.

MEASURING RADIOACTIVITY

Radiation is invisible, but it is possible to measure radioactivity using a Geiger counter. A Geiger counter contains a Geiger–Müller tube connected to a detector or counter device. As the tube absorbs the radiation it transmits an electrical pulse to the detector/counter device. This produces a clicking sound or displays a count rate. A greater frequency of clicks, or a higher count rate, indicates the presence of higher levels of radiation.

People who work with radiation need to monitor radioactivity to ensure that they are not exposed to harmful levels.

They can monitor their levels of exposure by wearing a radiation badge.

Inside the badge there are different materials through which radiation can travel to different degrees and a piece of photographic film that absorbs radiation. The more radiation it absorbs, then the darker it becomes in colour when developed. The colour of the developed film and the degree of penetration of the different materials can be used to monitor the levels of radiation.

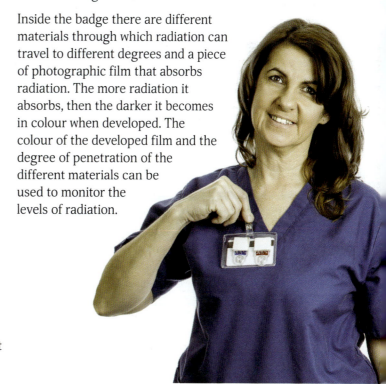

THINGS TO DO AND THINK ABOUT

1. Draw an atomic structure diagram for (i) hydrogen and (ii) helium.
   (Hint: atomic structure diagrams were covered in Unit 1)

2. Copy and complete the following table to show natural and artificial sources of background radiation.

Background radiation	
Natural sources	Artificial sources

CHEMICAL ANALYSIS

Chemical analysis is involved in all aspects of chemistry. It is important that you understand the significance of analysis and can carry out the following simple analytical techniques.

PAPER CHROMATOGRAPHY

Paper chromatography is a simple separation technique that can be used to analyse mixtures containing coloured compounds such as inks and dyes.

The term chromatography means a 'drawing of colour' originating from the Greek words chromato (colour) and graph (drawn/written).

How does it work?

Inks and dyes are not pure substances but are mixtures of coloured compounds. Each of these individual compounds has a different solubility.

Analysing black ink

A sample of the black ink to be analysed is 'spotted' onto a piece of absorbent paper (usually filter paper).

This paper is then placed in a suitable solvent depending on the solubility properties of the ink, for example water or ethanol.

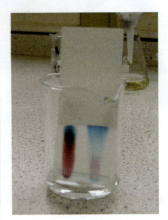

As the paper absorbs the solvent it will dissolve the individual constituent compounds of the ink, which are then carried up the paper by the solvent.

The more soluble a compound is, the further it will travel up the paper.

The separated ink produces a **chromatogram**, which can be analysed.

Analysing black ink by chromatograph

Samples of known compounds for comparison can also be separated by chromatography and a direct comparison can then be made between the compounds found in the ink and the known compounds.

OTHER SEPARATION TECHNIQUES

Separation techniques including filtration, evaporation and distillation will be covered in the skills section of this study guide.

DON'T FORGET

The solubility of a substance is the degree to which that substance (solute) will dissolve in a given solvent.

DON'T FORGET

Ethanol is the chemical name for common alcohol.

Flame tests

All ionic compounds contain metal and non-metal ions.

When ionic compounds are burned in a Bunsen flame the metal ion present in the compound produces a characteristic flame colour.

The flame colour produced can be used to identify the metal ion found in the compound.

Carrying out a flame test

A loop of metal wire is dipped into a sample of the compound being tested.

The sample is burned in the half-and-half Bunsen flame.

The colour produced is recorded and can then be matched to the known flame colours produced by different metal ions.

This allows indentification of the metal ion found in the compound.

Metal	Ion	Flame colour
barium	Ba^{2+}	green
calcium	Ca^{2+}	orange–red
copper	Cu^{2+}	blue–green
lithium	Li^+	red
potassium	K^+	lilac
sodium	Na^+	yellow
strontium	Sr^{2+}	red

Flame colours table

Using precipitation reactions

A precipitation reaction takes place when two soluble ionic solutions (aqueous) are added together.

As the ions from the two solutions mix together, one of the metal ions present and one of the non-metal ions present react together to form an insoluble salt which precipitates (solidifies) out of the solution.

Precipitation reactions can be used to identify the type of ions present in a sample.

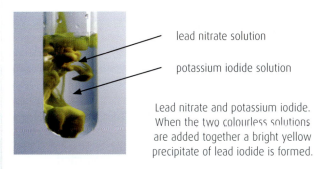

lead nitrate solution

potassium iodide solution

Lead nitrate and potassium iodide. When the two colourless solutions are added together a bright yellow precipitate of lead iodide is formed.

The word equation for this reaction is:

lead nitrate (aq) + potassium iodide (aq) → lead iodide (s) + potassium nitrate (aq)

pH TESTING

pH testing can be used to monitor the environment. Water and soil samples can be tested using pH indicators to determine whether they are acidic, alkaline or neutral.

universal indicator

Testing water

Indicators can be added directly to water samples to determine the pH.

The pH indicator is green and when it is added to the water sample in the test tube it changes colour and turns orange.

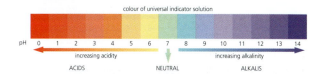

colour of universal indicator solution

pH 0 1 2 3 4 5 6 7 8 9 10 11 12 13 14

increasing acidity   increasing alkalinity

ACIDS   NEUTRAL   ALKALIS

Matching this to the universal (pH) indicator chart above we can see that orange indicates a pH value of 4. The water sample has a pH value of 4.

This means that the water sample is acidic because a pH value less than 7 indicates an acidic solution.

Testing soil

A soil sample must first have pure water added to it to dissolve any compounds that are present.

Step 1 – pure water is added to soil sample in a beaker.

Step 2 – the soil/water mixture is filtered to produce a colourless solution in a test tube.

Step 3 – universal indicator is added to the test tube.

The indicator is added and the colour change produced is matched to the pH chart to identify the pH value to determine whether the soil sample is acidic, neutral or alkaline.

Using a pH meter

A pH meter is an electronic device that measures pH.

The meter contains a probe that detects the pH and displays a digital reading of the pH value.

THINGS TO DO AND THINK ABOUT

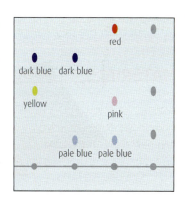

red

dark blue dark blue

yellow

pink

pale blue pale blue

1. Chromatography was used to identify the different coloured substances in an ink sample taken from a black felt-tipped pen. A spot of ink from the felt-tip pen was dabbed onto a piece of filter paper. Dots of the following comparison inks were also added: green ink, blue ink and purple ink. Water was used as the solvent. The chromatogram shown to the right was produced.

    (a) Which colours were present in the ink from the black felt-tip pen?
    (b) Suggest why a pencil was used to draw the line on the paper.

2. A chloride salt was tested by heating it in a Bunsen flame. The salt produced a lilac flame on burning. Name the metal present in the salt.

3. Grit spread on icy roads in winter contains a mixture of sand and sodium chloride, a soluble salt.
    (a) Suggest a method you could use to extract the sodium chloride from the grit that would produce a solid sample for analysing.
    (b) Explain how you could analyse the sodium chloride sample to (i) test its pH value and (ii) prove it was sodium chloride.

KEY AREA QUESTIONS

METALS AND ALLOYS

1.

The Kelpies are 30 m high stainless steel structures found next to the Forth and Clyde canal in Falkirk.

Stainless steel is an alloy made from iron, carbon and chromium.

(a) What is meant by the term alloy?

(b) Suggest two reasons why a stainless steel alloy was used as the material to make the Kelpies rather than iron metal.

2. Iron metal can be coated in a physical barrier to prevent corrosion.

(a) What common name is given to the corrosion of iron?

(b) Why does a physical barrier protect iron from corrosion?

(c) What metal is used to galvanise a piece of iron?

(d) Apart from galvanising, give two other methods used in the physical protection of iron.

MATERIALS

1.

Pop-up tents are made from a plastic called polyester. Polyester is a thermosoftening plastic.

What does the term thermosoftening mean?

2.

The material used to make some plastic food containers is called poly(propene).

(a) Name the monomer used to make poly(propene).

(b) Give two advantages of using plastic to make food containers.

(c) Why is burning not a suitable method of disposing of plastics?

3.

Polystyrene is a plastic used as a material for packaging.

(a) Polystyrene is a polymer. What is a polymer?

(b) Name the type of chemical reaction used to make polystyrene.

(c) Biodegradable packaging chips made from natural materials such as cornstarch are now being used instead of polystyrene. Suggest why.

FERTILISERS

1. As the population of the world increases, more crops are needed to meet the growing demand for food.

   Fertilisers supply essential elements needed for healthy plant growth.

   (a) Name the three essential elements needed for healthy plant growth.

   (b) Fertilisers can be natural or synthetic.

      (i) Give an example of a natural fertiliser.

      (ii) What is meant by the term synthetic?

   (c) Ammonium nitrate (NH_4NO_3) is a chemical compound found in synthetic fertilisers.

   Suggest two reasons why ammonium nitrate is suitable for use as a fertiliser.

   (You may wish to use page 8 of a National 5 data booklet to help you)

   (d) Calculate the percentage composition of nitrogen in ammonium nitrate.

NUCLEAR CHEMISTRY

1. Background radiation is all around us. This radiation can be produced from natural or artificial sources.

   X-rays used in medicine, cosmic rays, nuclear power stations, radon gas trapped in rocks, smoke detectors, food and drink, living plants and nuclear weapons are all sources of background radiation.

   From the list above:

   (a) Give two natural sources of background radiation.

   (b) Give two artificial sources of background radiation.

   (c) What is the name of the instrument that can be used to measure radioactivity?

CHEMICAL ANALYSIS

1. The table below shows the optimum (best) pH of soil for growing different vegetables.

Vegetable	Optimum (best) pH of soil
Carrots	5.5–6.5
Cauliflower	5.5–7.5
Potatoes	4.5–6.0
Cabbage	6.0–7.5

   A gardener wants to grow vegetables in their back garden.

   Describe the method you could use to analyse a sample of soil collected from the back garden to determine which vegetable would be the best one to grow.

2. Chromatography can be used to analyse the food dyes used to colour sweets.

   The chromatogram below was produced when a brand of sweets were analysed to see if they contained the food dye Blueberry blue.

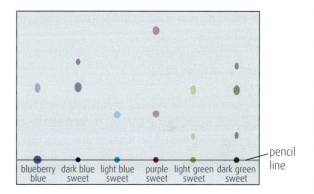

   (a) Which sweets contain the Blueberry blue dye?

   (b) Explain how you used the chromatogram to answer part (a).

EVERYDAY LABORATORY SKILLS 1

One of the important parts of this course is the development of key scientific skills. As well as important practical skills, you should be given opportunities to develop problem-solving and data-handling skills. Practical skills are an important part of any science subject. In chemistry there are some basic practical techniques that are regularly used and so it is important for you to be able to carry them out correctly. Not all of the practical skills outlined in the course are dealt with here. In this guide, you have already encountered methods for measuring pH, testing for starch, testing for sugars and testing for unsaturation, following rates of reaction, electrical conductivity and cells, and flame tests.

USING A BALANCE

A balance is an essential piece of equipment in any laboratory. It is used to measure mass (how much of something there is). In the lab, the unit for mass is usually grams, g.

Some balances are used for measuring an approximate mass of a substance and will have only one decimal place. Others are used for more accurate measurements and may have four decimal places – these are called analytical balances.

Here are some important instructions to follow when using a balance.

- Ensure that the pan is clean. If the pan is dirty, then it will not be possible to take an accurate mass measurement.
- Switch on the balance using the power button.
- Wait for the display to settle. When it is ready for use it will have a reading of zero grams.
- Gently place the object being measured onto the centre of the pan. Record the reading, with units, being sure to record all the decimal places. For example, a reading of '1·50 g' would be written down as 1·50 g and not as 1·5 g.

There are two main methods used when weighing a substance.

DON'T FORGET

Different balances will display zero differently. A one-decimal-place balance will show 0·0 g whereas a two-decimal-place balance will show 0·00 g.

DON'T FORGET

Make sure that all the decimal places on the display of the balance are recorded. A balance reading of 1·30 g should be recorded in this way and should not be simplified to 1·3 g.

1. Using the TARE (or ZERO) button

- Switch on the balance, wait for the display to settle and ensure the display reads zero grams.
- Place the beaker (or whatever is being used to contain the substance) on the balance.
- Press the TARE (or ZERO) button and wait for the display to read zero.
- Add an approximate mass of the substance.
- In your notes, record the reading on the balance.

2. Weighing by difference.

- Switch on the balance, wait for the display to settle and ensure the display reads zero grams.
- Place the beaker (or whatever is being used to contain the substance being weighed) on the balance.
- In your notes, record the reading on the balance (Reading A).
- Add the substance being weighed.
- In your notes, record the new reading on the balance (Reading B).
- The mass of the substance can then be calculated by B – A.

DISTILLATION

Distillation is a common technique used to separate a mixture when at least one of the substances in the mixture is a liquid. In the diagram, the mixture being separated is ethanol and water.

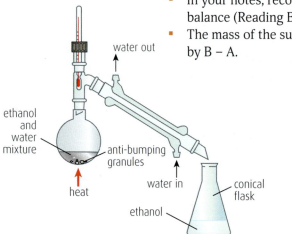

water out

ethanol and water mixture

anti-bumping granules

heat

water in

conical flask

ethanol

Ethanol and water have different boiling temperatures. The boiling point of ethanol is 78°C and that of water is 100°C. Heat is supplied to the mixture until the ethanol begins to boil and turns into a gas (**evaporates**). As the ethanol vapour passes the thermometer, the temperature measured will correspond to the temperature of ethanol vapour (around 78°C). The gas then enters the **condenser** and **condenses** back into a liquid, drips down the condenser and collects in the conical flask. This liquid is called the **distillate**. In this experiment, the distillate will be ethanol.

 **DON'T FORGET**

The condenser consists of an inner tube, through which the condensing gas passes, and an outer jacket. The outer jacket is connected to a tap (water is passed in at the bottom of the condenser and out at the top) and a small flow of cold water is passed through the jacket. This keeps the condenser cold and allows the gases to be condensed.

METHODS OF HEATING

In the laboratory, there are a number of methods that can be used to provide heat to a process or reaction.

The most common method of heating uses a Bunsen burner; this provides a quick and easy to control source of heating and is suitable for most school experiments.

When flammable substances are being used it is not advisable to use a Bunsen burner as the flame can cause the substance to ignite. In cases like these, sometimes boiling water from a kettle will provide enough heat.

In the distillation of ethanol, a **heating mantle** may be used to provide the heat needed to boil the mixture. A heating mantle is an electric heater specially designed to hold the round-bottomed flask holding the

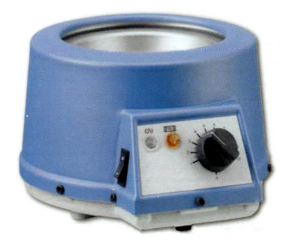

An electric heating mantle – often used to heat mixtures during distillation.

ethanol and water mixture. In this case, it is not safe to use a Bunsen burner as ethanol is flammable.

A hot-plate

An alternative method of heating a flammable substance is to use an electric hot-plate. This is similar to an electric hob that you may have on your cooker at home.

THINGS TO DO AND THINK ABOUT

1. A student was weighing calcium carbonate for use in an experiment. The following results were noted.
   Mass of beaker = 109·22 g
   Mass of beaker + calcium carbonate = 114·10 g
   (a) Calculate the mass of calcium carbonate used by the student.
   (b) Describe an alternative method for weighing out this mass of calcium carbonate.
2. State the meaning of:
   (a) distillation, (b) evaporate, (c) condenser, (d) distillate.
3. The mixtures below are to be separated by distillation. Which mixtures could be safely separated using a Bunsen burner?
   (a) salty water or (b) cyclohexane/hexane?

EVERYDAY LABORATORY SKILLS 2

COLLECTING A GAS

Collecting a gas produced in a chemical reaction is a useful technique. It can allow the gas products to be identified and the volume of the gas to be measured. This allows information to be calculated about the reaction quantities. There are different techniques used to collect a gas depending on its properties.

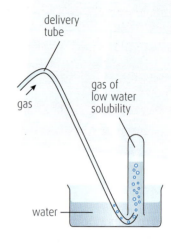

Collecting a gas by displacing water

This method is one of the most common used in school laboratories and works well for gases that are not very soluble in water, such as carbon dioxide and hydrogen. It is also often used to collect oxygen.

The gas produced in the reaction travels through the delivery tube, where it rises to the top of the test tube. This causes the water in the test tube to be pushed down and out at the bottom.

Collecting gases that are less dense than air

This method allows gases that are less dense than air (for example, ammonia) to be collected, as well as providing a way to collect dry hydrogen.

The gas produced in the reaction travels through the delivery tube, where it rises to the top of the test tube. This causes the air in the test tube to be pushed down and out at the bottom.

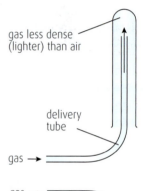

Collecting gases that are more dense than air

This method allows gases that are more dense than air (for example, carbon dioxide and chlorine) to be collected.

The gas produced in the reaction travels through the delivery tube where it sinks to the bottom of the test tube. This causes the air in the test tube to be pushed up and out of the top.

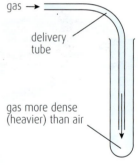

Measuring the volume of gas produced

In some experiments it may be necessary to measure the volume of gas produced in the reaction. This can be useful when measuring the rate of a reaction.

This can be done with a gas syringe; the gas produced in the reaction pushes the plunger and the volume of gas collected can be measured by reading the position of the plunger against the graduations.

A more common method in the school laboratory is to use an upturned measuring cylinder and displace the water from the measuring cylinder.

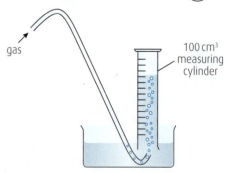

TESTING GASES

Testing for gases is an important technique to learn in chemistry. A chemical test is a procedure that can be used to prove the identity of a substance. This is because a chemical test will only give a certain result for that one substance.

There are a number of gas tests that you need to know about.

Testing for oxygen

Oxygen gas will relight a glowing wooden splint.

Testing for hydrogen

Hydrogen gas burns with a squeaky pop.

Testing for carbon dioxide

To test for carbon dioxide, the test gas is mixed with limewater (calcium hydroxide solution). Limewater turns cloudy (or milky) when carbon dioxide is present.

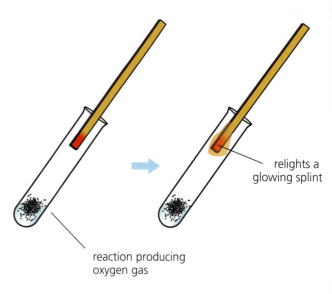

relights a glowing splint

reaction producing oxygen gas

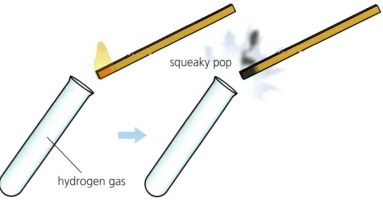

squeaky pop

hydrogen gas

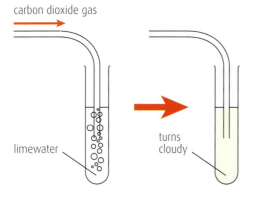

carbon dioxide gas

limewater

turns cloudy

DON'T FORGET

A lit splint will be extinguished if it is put into a test tube of carbon dioxide gas. This cannot be used as a test for carbon dioxide as other gases (nitrogen, for example) will also cause a lit splint to be extinguished.

THINGS TO DO AND THINK ABOUT

1. In each of the examples, describe the most appropriate method of collecting the gas.

   (a) Sulfur dioxide is much denser than air and is extremely soluble in water.

   (b) Carbon monoxide is insoluble in water and is slightly less dense than air.

   (c) Nitrogen dioxide is a very dense, brown gas that dissolves in water, forming an acid.

   (d) Helium gas is much less dense than air and is used to fill party balloons. The gas is very unreactive. A sample is required that does not contain any trace of water.

2. A student was given three test tubes. One test tube contained oxygen gas, another contained hydrogen gas and the third contained carbon dioxide gas. The labels had fallen off the tubes and so the student had no idea which gas was in each test tube. Describe how the student could show which gas was in each test tube.

MAKING A SALT

Salts are useful chemicals. As well as common salt (sodium chloride) that is added to fish and chips, salts are used in fertilisers, to make plaster casts and in indigestion remedies. There are different methods that can be used to make salts. The method chosen will depend on which salt is required.

MAKING A SALT BY NEUTRALISING AN ACID WITH A SOLUBLE BASE

During a titration, an acid is added from a burette to a soluble base (alkali) in a conical flask until neutralisation occurs. The end of the reaction is seen when an added indicator changes colour. The products in the conical flask will be salt and water. An example is the neutralisation of dilute hydrochloric acid using sodium hydroxide solution. The salt produced in this reaction is sodium chloride.

To obtain a pure sample of the salt, the water is evaporated. This can be carried out using a Bunsen burner and an evaporating basin. A Bunsen burner is a safe method of heating to use in this experiment as none of the reactants or products are flammable.

Care is taken during heating to make sure that not all the water is removed from the evaporating basin. The small volume of water remaining will evaporate naturally from the basin.

The salt will be left in the evaporating basin.

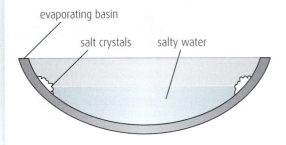

evaporating basin
salt crystals   salty water

salt solution   evaporating dish

DON'T FORGET

When carrying out a titration, concordant volumes are those that are within 0.2 cm³ of one another. 22.1 and 22.3 cm³ are concordant volumes, but a volume of 22.6 cm³ obtained in the same titration would not be concordant.

DON'T FORGET

A salt is a compound in which the hydrogen ion of an acid has been replaced by a metal ion (or sometimes an ammonium ion). In other words, it is a product of neutralising an acid.

DON'T FORGET

The second part of a salt name comes from the acid that has been neutralised. The first part of a salt name comes from the base that was used to neutralise the acid. Check out the section 'Neutralisation' on pp. 26–27 to revise this.

MAKING A SALT BY REACTING AN ACID WITH A METAL OR AN INSOLUBLE BASE

Salts can also be produced by adding a reactive metal to an acid. For example, magnesium sulfate can be made by adding magnesium to dilute sulfuric acid. This reaction also produces hydrogen gas:

magnesium + sulfuric acid → magnesium sulfate + hydrogen

The acid will be completely reacted when no more gas is produced and unreacted magnesium is left in the beaker. To obtain the salt on its own, the unreacted magnesium must first be removed by **filtration**.

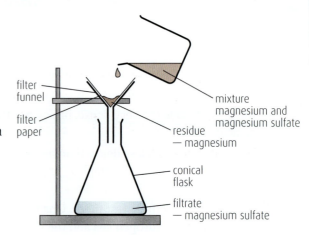

filter funnel
filter paper
mixture magnesium and magnesium sulfate
residue — magnesium
conical flask
filtrate — magnesium sulfate

The solid collected in the filter paper is called the **residue**. In this example the residue is unreacted magnesium. The liquid or solution collecting in the conical flask is the **filtrate**. In this example the filtrate is the magnesium sulfate solution.

The magnesium sulfate solution can then be poured into an evaporating basin and the water evaporated using a Bunsen burner.

This same process could be carried out using an insoluble salt such as magnesium carbonate or magnesium oxide.

MAKING AN INSOLUBLE SALT

The salts that were made in the previous experiments were all soluble in water and the water was evaporated to obtain the solid salt. Insoluble salts such as silver(I) chloride and lead(II) iodide can be made by reacting two soluble salts together. If we wanted to make pure silver(I) chloride we could do this by reacting a solution of silver(I) nitrate and a solution of sodium chloride.

Step 1	Step 2	Step 3	Step 4
a solution of silver(I) nitrate is added to a solution of sodium chloride	the precipitate is filtered off	the filtered precipitate is washed several times with deionised (pure) water	the silver(I) chloride is carefully scraped off the filter paper into a dish and dried in an oven

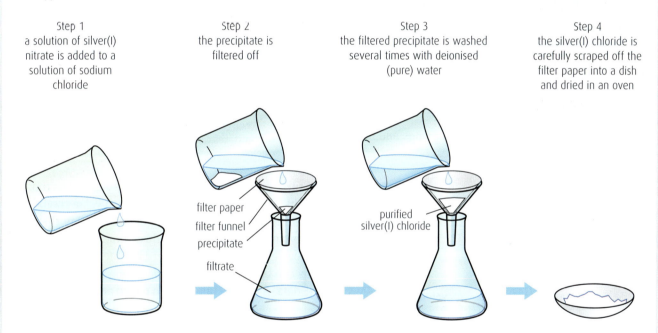

This process is known as **precipitation**. The insoluble salt formed in the reaction is removed by **filtration**.

THINGS TO DO AND THINK ABOUT

1. Name the salt produced when:

    (a) copper carbonate reacts with dilute sulfuric acid;

    (b) potassium hydroxide reacts with dilute nitric acid;

    (c) dilute hydrochloric acid is reacted with zinc.

2. State the meaning of

    (a) filtration;

    (b) filtrate;

    (c) residue.

3. Explain how the water could be removed from a solution of a salt made during a titration reaction.

TABLES AND BAR GRAPHS

Well-presented tables and graphs provide a concise summary of experimental data and are incredibly useful for presenting results.

CONSTRUCTING A TABLE OF EXPERIMENTAL DATA

There is rarely a correct method for constructing a table of experimental data, but there are some guidelines that apply to all table designs.

- Tables look best if they have a top, bottom and sides:

Table heading 1	Table heading 2

Table heading 1	Table heading 2

- Units should be included in the headings and not listed with each entry in the table.

Time	Volume
2 min	15 cm³
4 min	23 cm³
6 min	30 cm³

Time (mins)	Volume (cm³)
2	15
4	23
6	30

- Select the best unit for your experimental data.

Sometimes the data you have may contain a mixture of units. A common example is when recording time. The experiment may be collecting data every 10 seconds and if the experiment lasts for several minutes it may be that the time noted is a mixture of seconds and minutes. It is important to select an appropriate unit – in this case either seconds or minutes. Only one type of unit should be used.

CALCULATING MEANS (AVERAGES)

To improve the reliability of results in chemistry, experiments should always be repeated. Often a table is used to show data from repeat experiments.

For example, a student carried out an experiment to find the time taken for a reaction to finish. The experiment was repeated another two times. The following results were obtained.

	Experiment 1	Experiment 2	Experiment 3
Time taken (seconds)	320	315	322

All three experiments gave similar results and so the results are said to be reliable, but which value should the student use? The answer is, of course, all of them. Carrying out the experiment three times has shown that the correct time lies somewhere between 315 and 322. This can be represented by calculating a mean value for the data. A mean value is often called an average.

To calculate an average, all the values are added together and the total is divided by the number of sets of values.

So, the average time for this experiment would be:

$$\frac{320 + 315 + 322}{3}$$

= 319 seconds.

Any repeat experiments which give results that are very different from the others can be disregarded as being rogue data because something went wrong during the experiment.

CONSTRUCTING A BAR GRAPH

Graphs are an incredibly useful tool to a scientist as they allow data to be displayed and understood in a quick and easy manner. They allow generalisations to be made and conclusions to be drawn for experiments. There are two types of graph commonly used: bar graphs and line graphs.

A bar graph is commonly used when only one set of experimental data involves numbers. For example, if an experiment was carried out to measure the different energy values of various fuels, this information would be represented by a bar graph. It is common to have the numerical values plotted on the y-axis.

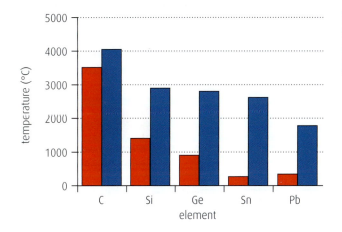

This bar graph shows group 4 elements on the x-axis with the numerical values for their melting and boiling points plotted on the y-axis.

Here are some general guidelines:

- The scale for the y-axis needs to be chosen carefully. It needs to be linear, for example, 0, 1000, 2000, 3000, 4000 and not 0, 500, 1000, 2000, 4000.
- The graph should be as big as possible and should certainly use at least half of the graph paper provided.
- Bars should be drawn using a pencil and ruler and should be of equal width.
- Both axes need to be carefully labelled, using units as appropriate.

THINGS TO DO AND THINK ABOUT

1. Suggest three improvements that could be made to the table below.

Acid used	Trial 1 volume collected	Trial 2 volume collected
hydrochloric acid	500 cm³	520 cm³
sulfuric acid	990 cm³	1 litre

2. The table shows the results for an experiment to find out which metal, when paired with zinc, gave the highest voltage in an electrochemical cell.

Electrode	Voltage (V)		
	Trial 1	Trial 2	Trial 3
iron	0.5	0.4	0.5
lead	1.2	1.1	1.3
copper	1.8	1.7	1.6
magnesium	0.8	0.8	0.7

   Calculate the mean (average) voltage for each metal.

3. Construct a bar graph to show the average voltage for each metal from the data in Question 2.

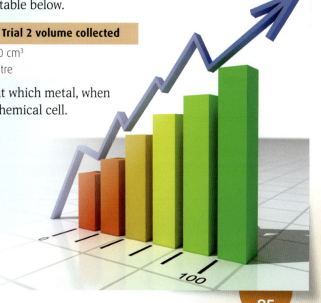

DRAWING A LINE GRAPH

CONSTRUCTING A LINE GRAPH

Line graphs are used to present data when both sets of experimental data involve numbers and these numbers are continuous.

Generalisations and conclusions can only be made from graphs if they have been correctly drawn.

We shall use the example of measuring the volume of carbon dioxide gas produced every 30 seconds when marble chips react with dilute hydrochloric acid. The following table of data was produced:

Time (s)	Volume (cm³)
0	0
30	12
60	32
90	51
120	64
150	73
180	80
210	81
240	81

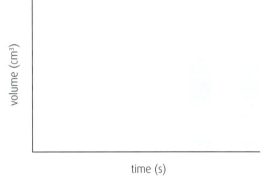

In most cases, the variable you are controlling is plotted along the x-axis (bottom). In this case the variable being controlled is time – we are choosing to measure the volume every 30 seconds. Volume would be plotted on the y-axis.

Scales for both the x- and y-axes need to be chosen carefully.

- The graph should be as big as possible and should certainly use at least half of the graph paper provided.

- The scales chosen need to be linear. For example, if one box on a graph is worth 10 cm³, then each box on that graph must equal 10 cm³. This may mean that sometimes a data point will be plotted between two divisions (if, for example, 15 cm³ was being plotted, this would come between the 10 cm³ and 20 cm³ divisions).

Using the graph paper below, scales can be drawn as shown.

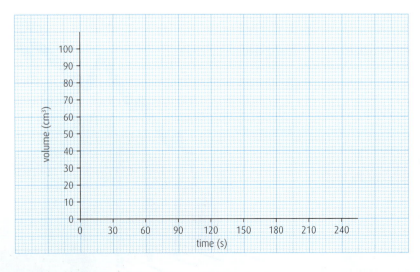

- The data points are plotted carefully using small marks – crosses are commonly used but small dots are also acceptable.

- You should also consider if your graph should be going through the origin. This will depend on the experiment. In the experiment above, at the start of the reaction the volume of gas would be zero and so when drawing the line it should go through the origin.

Your graph should now be ready for a line of best fit to be drawn.

DRAWING A LINE OF BEST FIT

When drawing a line on your graph it is necessary to first of all consider what your experimental data represent. Mostly, the data you have plotted on your graph will have been from measuring a quantity in a reaction. There will be errors associated with this value (is it easy to read an exact volume measurement of a gas at exactly 30 seconds?). As such the value you recorded (12 cm^3) will not be exactly this value and could be a little less or a little more than this value. For this reason, lines of best fit are drawn. This shows a general trend line that goes through, or as close to, as many data points as possible.

There are two main types of line of best fit.

Best-fit straight line

This is probably the easiest to draw. If your experimental data look to be in a straight line, then, using a ruler, a straight line can be drawn that takes in as many of the data points as possible. This is done 'by eye', and an easy way to do it is to try to have as many points below the line as above it.

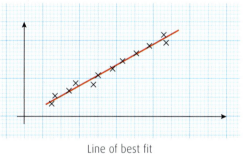

Line of best fit

Best-fit curve

This is slightly trickier as the line will need to be drawn free-hand. Again, the line should go through, or as close to, as many of the data points as possible and you should aim to have as many points below the line as above. The line needs to be smooth and sharp (do not try to sketch the line but draw it using one continuous hand movement).

The experiment outlined on page 86 will produce a similar-shaped graph to the best-fit curve shown here.

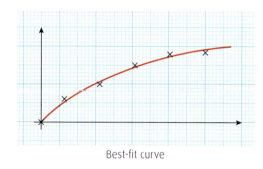

Best-fit curve

THINGS TO DO AND THINK ABOUT

1. For the following examples, decide which variable should be plotted along the *x*-axis.

(a) In an experiment to determine the rate of a reaction, the mass of the reaction was recorded every 2 minutes.

(b) A reaction was carried out at 10, 20, 30 and 40°C and the time taken for the reaction was recorded.

2. Look at this graph and explain what is wrong.

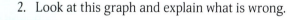

EXPERIMENTAL DESIGN

The ability to be able to design an experiment is a useful skill for a chemist to have. Without a good experiment design, the results cannot be relied upon.

FAIR TESTING

The first thing to consider when designing an investigation is: the investigation fair? A fair test is when only one of the factors (variables) of the experiment is altered at a time, whilst all the other variables are kept the same.

Let us consider an investigation to see how altering the solution used as an electrolyte affects the voltage produced in an electrochemical cell. The diagram below shows the first experiment carried out during this investigation.

To investigate what effect the solution used as an electrolyte will have on the voltage produced by this cell, the solution used as an electrolyte will need to be changed in further experiments. This is the variable that is to be investigated. To ensure that the changes in voltage are only due to the changes in the electrolyte, all other variables will need to be kept the same. This ensures that the investigation is fair. Copper and iron electrodes will need to be used, the volume and concentration of the electrolyte will need to stay the same and the temperature at which the experiment is carried out at will need to be constant.

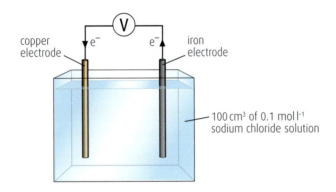

copper electrode   e⁻   e⁻   iron electrode

100 cm³ of 0.1 mol l⁻¹ sodium chloride solution

REDESIGNING AN EXPERIMENT

Sometimes when we look at our results we can see that the experiment needs to be redesigned in some way.

Sometimes results from experiments can be completely unexpected and this can mean that the experimental design was wrong.

EQUIPMENT DIAGRAMS

In designing an experiment, a diagram of the equipment is often required. There are some key points to consider when drawing such diagrams. First, break down the experiment into two key parts: reactants and products.

Reactants

Begin by considering how the reaction is going to take place:

- does the experiment need to be heated or cooled in ice?

- do any of the reactants need to be added slowly over a period of time?

Products

Next, consider the products – how are they to be collected?

Experiments involving collecting gases need to be sealed systems and the diagram must show this. Stoppers or bungs and delivery tubes are common in experiments involving gases.

When gases are being passed through test tubes, care needs to be taken to show correctly how the gas will be drawn through the equipment. This is shown in the following diagram.

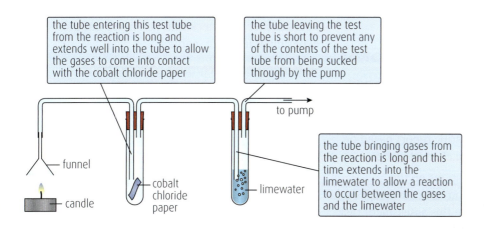

the tube entering this test tube from the reaction is long and extends well into the tube to allow the gases to come into contact with the cobalt chloride paper

the tube leaving the test tube is short to prevent any of the contents of the test tube from being sucked through by the pump

to pump

funnel

candle

cobalt chloride paper

limewater

the tube bringing gases from the reaction is long and this time extends into the limewater to allow a reaction to occur between the gases and the limewater

DRAWING EXPERIMENTAL DIAGRAMS

Chemists are often required to draw diagrams of the equipment or apparatus used in an experiment. The following guidelines should be followed:

- A ruler and pencil should be used.
- Labels are essential for equipment and chemicals.

THINGS TO DO AND THINK ABOUT

Eggshells are made up mainly of calcium carbonate. A pupil carried out an experiment to react eggshells with dilute hydrochloric acid. A gas was produced. Complete the diagram to show the apparatus that could have been used to measure the volume of gas produced.

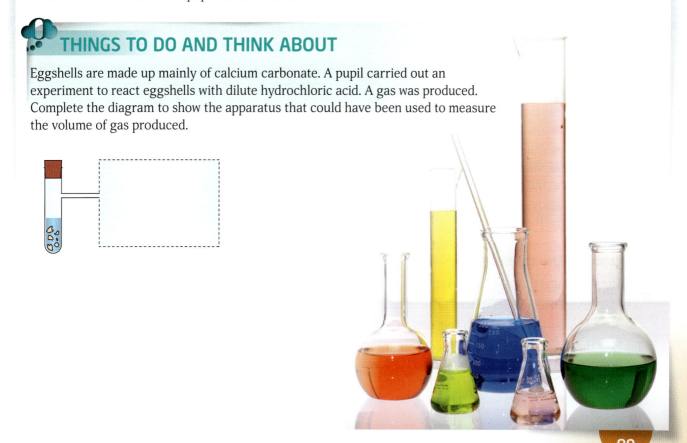

ASSESSMENT

UNIT ASSESSMENT

The good news is that you are not required to sit a final exam at the end of your National 4 Chemistry studies in order to achieve a course award. Instead of a final exam you are required to pass four assessment tasks. These four tasks will be taken in school and they will be marked by your teacher on a pass or fail basis.

WHAT ARE THE ASSESSMENT TASKS?

Assessment task 1 – Preparation of a scientific report on a chemistry experiment or practical investigation.

Assessment task 2 – A short scientific report based on research of a chemistry topic.

Assessment task 3 – A set of questions on the chemistry you have covered. You will complete this assessment task for each unit.

Assessment task 4 – This task is called the Added Value Unit or Assignment.

ASSESSMENT TASK ONE

To be successful in this assessment you must plan and carry out an experiment or practical investigation and then prepare a scientific report about the work undertaken.

Your report must cover the following areas:

1. Planning

This section of the report must include:

(a) The aim of the experiment – what you are trying to find out. Some examples of suitable aims are

  (i) To find out how the concentration of acid affects the volume of alkali needed to neutralise the acid.

  (ii) To find out how the temperature affects the rate of the reaction.

  (iii) To find out what happens to the pH of an alkali when it is diluted with water.

(b) The variable(s) that need to be kept constant.

(c) Measurements or observations to be made.

(d) The resources (apparatus and chemicals) to be used.

(e) How the experiment will be carried out – the procedure. This must be written in such a way that another National 4 pupil could use the information to repeat the experiment.

(f) Any safety precautions needed. This could include the use of safety glasses, not using naked flames to heat up flammable liquids, washing out apparatus before use and wiping up any spillages that happen.

2. Following procedures safely.

Your teacher will check that you carry out the experiment in a safe manner.

3. Making and recording observations/measurements accurately.

4. Present results in an appropriate manner

Some things you might include are:

  (i) Tables with appropriate headings and units.

  (ii) Calculations of average results from repeated experiments.

  (iii) Colour changes.

  (iv) Bar charts or line graphs. Charts or graphs should have appropriate scales, labels and units. The bars or points should be plotted correctly.

5. Drawing valid conclusions

Your conclusion must relate to your aim.

The valid conclusions for the aims set out above could be written as:

(i) The volume of alkali need to neutralise an acid increases as the concentration of the acid increases.

(ii) Increasing the temperature increases the rate of the reaction.

(iii) When an alkali is diluted with water, the pH of the alkali decreases.

6. Evaluating the procedure

In this final section of the report you must suggest something that would improve the experiment.

DON'T FORGET

A variable in an experiment is something which can be altered. For example; the volume of solutions used, the concentration of solutions used, the temperature and the type of solution used are all variables. For an experiment to be valid only one variable should be changed. Other variables must remain unchanged.

DON'T FORGET

Group work is allowed for this stage but you will need to show you have actively participated in any group work.

ADDED VALUE UNIT

ASSESSMENT TASK TWO

JUST A WEE NOTE
The impact can be positive and/or negative.

To be successful in this assessment you must select and research an application of chemistry from one of the key areas you have studied. The report should show the impact the application has on the environment or society.

When you have completed your report you should read over it and make sure that you have:
(i) stated the application of chemistry involved.
(ii) used appropriate chemistry knowledge to describe the application. For example, if your application was the use of thermoplastics for packaging, you could state that thermoplastics are plastics which melt and can therefore be moulded when they are heated.
(iii) stated how the environment and/or society has been affected by the application.
(iv) used appropriate chemistry knowledge to describe its effect.

For example, if your application was the use of fertilisers in farming, you could state that fertilisers have a negative impact on the environment because they are soluble in water and can therefore be washed out of the soil by rain, ending up in rivers causing pollution.

ASSIGNMENT

The assignment at N4 is in place of a final examination and is marked as a pass or fail.

It covers the following assessment standards:
1.1 Choosing, with justification, a relevant issue in chemistry
1.2 Researching the issue
1.3 Presenting appropriate information/data
1.4 Explaining the impact, in terms of the chemistry involved
1.5 Communicating the findings of the investigation
If you fail to pass one or more of these assessment standards, don't panic because you are allowed one resit opportunity.

You will apply your skills, subject knowledge and understanding to investigate a topical issue in chemistry and its impact on the environment and/or society. There are two stages to the assignment process:

Stage One – the research stage

First you must select a topical issue to base your assignment on.

DON'T FORGET
Remember to ask your teacher for the candidate guides for the assessment tasks.

Next you will gather information from a variety of sources (such as books/the internet/class notes/scientific journals) and this will allow you to create a candidate's log or journal.

Your teacher will help you with this stage, although the research must be your own work and you should record the sources for any information that you gather.

DON'T FORGET
Remember that Wikipedia is not a reliable source.

If you carry out an experiment(s) as part of the research stage, you may wish to include this work in your assignment.

Stage Two – the communication stage

In this stage you will produce your final assignment.

During the second stage of the assignment you will select, use and record as a minimum at least two appropriate sources from your research which will be included in your final communication.

DON'T FORGET
Remember to include the title and aim of any experiments.

JUST A WEE NOTE
This will take place under controlled conditions – this means your teacher will be present.

You must select, process and present information and/or data which you gathered in stage one relating to your chosen topical issue.

DON'T FORGET
Your teacher will support you throughout your Added Value Unit.

THINGS TO DO AND THINK ABOUT

You can choose how you communicate your findings and this may include one or more of the following:

- A written or word-processed report (200–400 words is the suggested length)
- A presentation, oral or digital which is accompanied

by supplementary material, for example presentation notes or presentation slides with notes.
- An information booklet/leaflet
- An information poster (this poster must include annotated notes).

ANSWERS

Reaction rates – p7

1. (a) All chemical reactions slow down as the temperature decreases. The reactions causing milk to "go off" will slow down if the milk is stored in a fridge.
   (b) Large particles have a smaller surface area than an equal mass of smaller particles and therefore they react more slowly. The same is true if large carrots are compared with smaller carrots.
   (c) The reaction is between the charcoal and oxygen. Blowing air onto the charcoal provides more oxygen causing the charcoal to glow more brightly.
2. 1 – Record mass of penny.
   2 – Place penny in sulfuric acid/zinc mixture and wait for reaction to stop.
   3 – Filter out penny and dry it.
   4 – Record mass of the penny. If it has not been used up its mass will be unchanged.
3. (a) F   (b) C and F   (c) A

Monitoring reaction rates – p9

1. (a) 15 cm^3 (b) 45 s   (c) 42 cm^3
   (d) Dotted should be steeper than the original line and it should become horizontal at 42 cm^3.

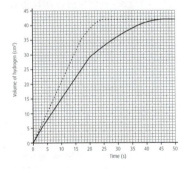

Atomic structure - the Periodic Table – p11

1. False, false, false, false, false, false, true, false, false, false.
2. Element 114 is flerovium and is named after the Flerov Laboratory of Nuclear Reactions of the Joint Institute for Nuclear Research in Dubna, Russia, where the element was discovered in 1998. The laboratory was named after the Russian physicist Georgiy Flerov. Element 116 is livermorium and is named after the Lawrence Livermore National Laboratory in California USA, which collaborated with the Joint Institute for Nuclear Research in Dubna, Russia to discover livermorium in 2000.

Atomic structure - the structure of an atom – p13

1. (a) 3   (b) 8   (c) Metal
   (d) An atom has the same number of protons and electrons. The number of positive charges are cancelled out by the number of negative charges.
2. (a) 15   (b) 31   (c) Phosphorus
3. 11 protons and 12 neutrons.

Names and formulae 1 – p15

1. (a) N_2O_5 (b) SF_6 (c) PCl_3
   (d) CS_2 (e) CO
2. (a) CH_4 (b) C_2H_6O (c) HBr
   (d) C_3H_6O (e) PBr_3
3. (a) Four   (b) 24
4. (a) HCN
   (b) Name ending 'IDE' usually indicates only two elements.

Names and formulae 2 – p17

1. (a) 1   (b) 4   (c) 1   (d) 2
   (e) 3   (f) 2
2. (a) LiBr   (b) Mg_3N_2 (c) SiO_2
   (d) $SrCl_2$ (e) Al_2O_3 (f) Na_2S
   (g) HgO   (h) $CrCl_3$ (i) $CoCl_2$
   (j) Mn_2O_5
3. (a) PCl_3 – does not break rules
   (b) CO – breaks valency rules
   (c) SO_2 – breaks valency rules
   (d) N_2O – breaks valency rules

Formula mass – p19

1. (a) 58.5   (b) 18   (c) 44
   (d) 95.5   (e) 102   (f) 40.5
   (g) 348   (h) 46   (i) 110
   (j) 431
2. (c)
3. (d)
4. (a)
5. (b)
6.

Compound	Formula mass
SO_2	64
CaO	56
H_2S	34
LiBr	87

Ionic and covalent bonding – p21

1. Ionic – magnesium chloride, lithium sulphate, potassium bromide, sodium nitrate, calcium oxide
   Covalent – nitrogen hydride, carbon chloride, hydrogen oxide

2. (a) A/D   (b) B/C   (c) B/C   (d) A
3. (a) sodium chloride, calcium fluoride
   (b) methane, carbon dioxide
   (c) copper, sulfur
   (d) sodium chloride, calcium fluoride
   (e) methane, carbon dioxide

Chemical reactions – p23

1. (a) 8.2°C   (b) X = 24.8°C Y = 20.2°C
   (c) Reactions 1 and 3 are exothermic – the temperature increases after the chemicals are mixed. In experiment 2 the temperature decreases after the chemicals are mixed. Reaction 2 is endothermic. Jennifer is **incorrect**.
2. (i) s   (ii) aq   (iii) aq   (iv) g
   (v) aq   (vi) aq   (vii) s   (viii) aq
3. (a) carbon monoxide + iron oxide → iron + carbon dioxide
   (b) magnesium + hydrochloric acid → magnesium chloride + hydrogen
   (c) copper chloride → copper + chlorine
4. (a) $SO_2 + O_2 \rightarrow SO_3$
   (b) $N_2 + H_2 \rightarrow NH_3$
   (c) $Mg + Cl_2 \rightarrow MgCl_2$

Acids and bases – p25

1. (a) vinegar, lemonade
   (b) sodium chloride
   (c) baking soda

Neutralisation – p27

1. (a) sodium nitrate + water
   (b) lithium chloride + carbon dioxide + water
   (c) potassium sulfate + carbon dioxide + water
   (d) zinc chloride + water
2. (b) and (d)
3. (a) Any number less than 7
   (b) 7
   (c) Sodium sulfate
   (d) sodium hydroxide + sulfuric acid → sodium sulfate + water
   (e) pH indicator, pH paper, universal indicator

Chemical changes and structure: key area questions – pp28-9

Rates of reaction:
1. (a) Measuring cylinder, gas syringe
   (b) 40 cm^3 (c) Experiment B
   (d) Measure the mass of gas lost as time passes. Measure the pH of the solution as time passes
2. Experiment 2 – line is steeper and goes horizontal at half the height of the original

Experiment 3 – line is steeper than other two lines and goes horizontal at twice the height of the original

3. (a) Concentration of acid is decreasing
    (b) 0.6 mol l^{-1}
    (c) It has not levelled out

Atomic structure:

1. (a) aluminium + oxygen → aluminium oxide
    (b) Al_2O_3
    (c) Ionic
    (d) Aluminium: mass number 27, number of protons 13. Oxygen: atomic number 8, number of neutrons 8.
2. (a) Covalent
    (b) C_4H_{10}
    (c) 44
    (d) 35-55°C
    (e) methane + oxygen → carbon dioxide + water
3. (a) Nucleus
    (b) Becuase is has an equal number of protons and electrons. Positive charge is cancelled out by the negative charge.
    (c) 11

Energy changes of chemical reactions:

1. (a) Temperature will increase
    (b) Endothermic

Acids and bases:

1. (a) Neutralisation
    (b) Acid rain: rivers polluted, trees/plants killed
    (c) (i) 7
        (ii) Universal indicator, pH indicator, pH paper
        (iii) Any number greater than 7
        (iv) Calcium carbonate is insoluble in water
2. (a) Nitric acid
    (b) potassium hydroxide + nitric acid → potassium nitrate + water
    (c) By evaporating the water

Fuels – p31

1. Peat is produced in the first stage of forming coal. It is partially decomposed vegetable matter that has been buried under layers of soil.
2. Sun and wind are needed to dry the peat in the bogs and so a ditch is dug to allow water to drain. The peat is sometimes extracted manually by cutting into peat bogs to produce peat 'bricks' which are left out to dry. It can also be harvested using vacuum harvesters.
3. Peat is similar to coal as it is formed from decomposed plants.
4. Peat is sometimes called a fossil fuel as it takes thousands of years to form and is the remains of dead plants. Hoever, some people do not regard it as a fossil fuel as it does not take as long as coal and oil to form.

Burning of fossil fuels – p33

1. (a) Magnesium + oxygen → magnesium oxide
    (b) Iron + oxygen → iron oxide
    (c) Sulfur + oxygen → sulfur oxide (or sulfur dioxide)
    (d) Carbon hydride → carbon dioxide + hydrogen oxide (water)
2. (a) The chemical word for burning.
    (b) When a substance combines with oxygen.
    (c) Reactions that give out energy.
    (d) Reactions that take in energy.

Combustion – p35

1. That the total mass of all products at the end of a reaction will be exactly the same as the total mass of all reactants at the start of the reaction.
2. It increases at the end of the reaction because it has combined with oxygen in the air to become iron oxide. It now contains the same number of iron atoms as it did at the start of the reaction, but it also contains the oxygen atoms that reacted with the iron. These oxygen atoms also have a mass and so the iron oxide product is heavier than the iron wool reactant.
3. Some combustion reactions produce a gas as a product and this gas may escape from the reaction.
4. Methane.
5. Methane + oxygen → carbon dioxide + water.
6.
7. To put out a fire, oxygen could be removed (by smothering the fire), heat could be removed (by adding water) or the fuel could be removed.

Crude oil p37

1. A liquid with a thick sticky consistency.
2. Many different compounds.
3. Plastics, dyes and medicines are three examples.
4. Fractional distillation.
5. Fractions.
6. Refinery gas, petrol, naphtha, kerosene, diesel oil and residue.
7. Their different boiling points.
8. Refinery gas.
9. Bigger molecules are harder to evaporate.
10. The residue is the most viscous.

Impact on the environment – p39

1. A natural process that helps to keep the Earth from being too cold. Greenhouse gases form a layer round the Earth like a blanket. These gases trap energy from the Earth and help to keep it warm.
2. The Earth would be too cold and nothing would survive.
3. Carbon dioxide and methane greenhouse gases.
4. The name given to the rapidly changing weather and temperatures that affect the whole planet.
5. Increased amounts of greenhouse gases in the atmosphere.
6. Acid rain is rain that has a pH below 7 due to the chemicals dissolved in it.
7. Nitrogen oxides – formed in car engines when the spark from the spark plug causes nitrogen and oxygen to react. They are also formed naturally by lightning.
8. Sulfur dioxide.
9. Sulfur impurities in fossil fuels react with oxygen. It is also formed naturally when volcanoes erupt.
10. Forests and rocks/cliff faces have been affected, damage to cars, buildings, statues, rail tracks and bridges.

Protecting the environment – p41

1. Wind power, solar power and hydroelectric power.
2. New fuels that do not produce carbon dioxide as well as methods to remove carbon dioxide from the atmosphere when a fossil fuel is burned.
3. Catalytic converters in cars help to reduce acid rain.

Biomass and biofuels – p43

1. Substances made from living or recently living organisms.
2. Wood is an example.
3. A fuel made from biomass.
4.

Biofuel	Source	Uses
Biomethane	Rotting rubbish	Same as natural gas – domestic heating, cooking etc.
Bioethanol	Fermenting sugar	Fuel for cars
Biodiesel	Vegetable oil	Fuel for cars

5. The crops that are grown to produce biomass fuels take in carbon dioxide and when these fuels are burned they give out carbon dioxide. Overall the quantity of carbon dioxide remains the same.

Hydrocarbon compounds – alkanes – p45

1. Fossil fuels.
2. One that has all its carbon atoms joined together by single covalent bonds.

3. A molecular formula is the simplest formula of a compound showing the symbol and number of each type of atom present. A full structural formula shows how all the atoms and bonds in the compound are arranged.
4. Pent-
5. Propane
6. Heptane
7. Cracking is a reaction in which bonds in an alkane are broken producing smaller molecules.
8. Aluminium oxide.
9. Alkanes and alkenes.
10. Alkanes → smaller alkanes + smaller alkenes.

Hydrocarbon compounds – alkenes – p47

1. By cracking crude oil fractions.
2. Ethanol and polymers (plastics).
3. Alkenes contain a carbon-to-carbon double bond; alkanes only contain carbon-to-carbon single bonds.
4. But-
5. That it contains a carbon-to-carbon double bond.
6. Heptene.
7. Hex-1-ene
8. Pent-2-ene
9. An alkene will rapidly decolourise bromine water.
10. (a) Bromine water would remain yellow–orange with butane.
    (b) Bromine water would rapidly decolourise with butene.

Everyday consumer products – p49

1. Photosynthesis is the process that plants use to capture energy from the Sun and turn it into glucose.
2. Carbon dioxide + water → glucose + oxygen
3. Starch, oils and proteins.
4. Carbon, hydrogen and oxygen.
5. $C_6H_{12}O_6$
6. Sugar can be reacted with concentrated sulfuric acid. This produces water and black carbon.
7. Hydrogen 2 to oxygen 1
8. Starch is a complex carbohydrate.
9. To store energy.
10. Respiration.
11. Glucose will react with hot Benedict's solution, turning it from a blue solution to a brick red cloudy solution.
12. Starch can be tested for using iodine solution, which will turn from a orangey brown colour to blue–black.

Alcohol – p51

1. Glucose → ethanol + carbon dioxide
2. Yeast.
3. The gas produced can be bubbled through lime water which will turn cloudy if the gas is carbon dioxide.

4.

Plant used in fermentation	Alcoholic drink
Grapes	Wine
Barley and hops	Beer
Potatoes	Vodka
Barley	Whisky

5. Yeast is killed when the alcohol concentration reaches 14%.
6. Distillation.
7. Ethanol has a lower boiling point than water so when an ethanol and water mixture is heated the ethanol evaporates first. The ethanol gas is cooled down in a condenser and collected as a liquid. The water remains as a liquid.
8. 10 cm^3 of pure alcohol.

Plants to products – p53

1. Example answer:

	Medicine 1
Name of medicine	Aspirin
Name of plant	Bark of willow and spirea flowers
Where is the plant found?	China, Europe, North America and Asia
Active ingredient	Salicylic acid
Role of chemists	Isolating salicin and modifying it
Uses and applications	Pain and fever reduction
Benefits to everyday life	Reduced deaths due to fever

2. Glycerin comes from vegetable oils and so from plants that are used to produce oils such as olives, trees and sunflowers. It is used in shampoos and cosmetics to hold in water.
3. Nettle leaves, gorse flowers, raspberries and blueberries.

Nature's chemistry: Key area questions pp54—55

Fuels:
1. (a) The fuels, like fossils, are formed from the remains of living things.
   (b) They are running out or burning them causes pollution.
   (c) That energy is given out.
   (d) (i) Petrol
       (ii) 45 kg
       (iii) 2 160 000 kJ
2. (a) Hydrogen and carbon
   (b) hydrocarbon + oxygen → carbon dioxide + water
   (c) Carbon dioxide turns lime water cloudy.
   (d) Carbon monoxide
   (e) Biofuel
3. Biodiesel or bioethanol used as fuels in cars, or biomethane used for heating and cooking.

Hydrocarbons:
1. (a) Boiling point
   (b) Refinery gas
   (c) Road surface, fuel for ships or lubricating oil
   (d) Petrol
   (e) Cracking
2. (a) Butane
   (b) C_3H_8
   (c) 153–154°C
   (d) Pentane
   (e) Alkenes rapidly decolourise bromine water.

Everyday consumer products:
1. (a) Glucose
   (b) Energy storage
   (c) Starch turns iodine solution blue–black
   (d) blue → brick red
   (e) Digestion
   (f) Starch is too big to pass through the gut wall.
2. (a) Carbon dioxide
   (b) Yeast
   (c) Distillation
   (d) Units
   (e) 82.5 cm^3

Plants to products:
1. (a) Willow bark or spiraea
   (b) Aspirin does not irritate the mouth, gut or stomach
   (c) Dyes/cosmetics/paper

Metals – p56

1. (a) Gold
   (b) Copper
   (c) Magnesium
   (d) Nickel
   (e) Potassium
2. (a) Zn
   (b) Ag
   (c) Al
   (d) Ca
   (e) Fe
3. (a) Fences, gates, bridges, railway tracks, car body parts, cutlery, etc.
   (b) Bridges*, railway tracks*, car body parts*
4. (a) Structures, fire grates - strong, high melting point
   (b) Jewellery, coins – malleable, shiny, unreactive
   (c) Bicycle frames, bodies of aeroplanes – strong, light, does not corrode
   (d) Thermometers – liquid at room temperature
   (e) Electrical cables – ductile, conducts electricity, can be stretched into wires

Reaction of metals – p59

1. Sodium is near the top of the reactivity series and is very reactive in water, copper is near the bottom, copper does not react with water.
2. iron + oxygen → iron oxide

3. K, Na, Li, Ca, Mg, Al, Zn, Fe, Sn, Pb, Cu, Ag, Au
4. (a) Z, X, Y
   (b) X – any from aluminium, zinc, iron, tin or lead
   Y – copper, silver or gold
   Z – potassium, sodium, lithium, calcium or magnesium

Extraction of metals – p61

1.

Metal	Method of extraction from ore
potassium sodium lithium calcium magnesium aluminium zinc iron tin lead copper silver gold	electrolysis
	heating in the presence of carbon or carbon monoxide
	heating oxide alone

2. That platinum is at the lower end of the reactivity series, has a lower reactivity.

3.

Name of ore	Chemical name	Chemical formula
Haematite	Iron oxide	Fe_2O_3
Bauxite	Aluminium oxide	Al_2O_3
Iron pyrite (fool's gold)	Iron sulfide	FeS_2

Corrosion of metals – p63

1. A chemical reaction in which a metal reacts with water and oxygen from the air to form a compound of the metal.
2. (a) Painting, galvanising, greasing moving parts
   (b) Magnesium or zinc could be used; both are higher up the reactivity series and will corrode before the iron.
   (c) Tin would not be suitable, because it is lower in the reactivity series and would cause the iron to corrode quicker.
3. Greasing a bicycle chain (P), galvanising an iron bin (P), tin plating (P), zinc bolted to ship's rudder (C).

Making electricity – p65

1. Magnesium and copper (further apart in the electrochemical series).
2. (a) (i) Voltage would be lower;
       (ii) there would be no voltage.
   (b) (i) Electron flow would stay in the same direction;
       (ii) electron flow would stop.

3. Zinc chloride + magnesium → magnesium chloride + zinc
4. (a) Lithium, potassium, calcium, sodium, magnesium or aluminium
   aluminium + zinc sulphate → aluminium sulphate + zinc
   (b) Iron, nickel, tin, lead, copper, silver, mercury or gold; they are all below zinc in the electrochemical series and so would not displace zinc.

Plastics – p67

1. Tetrafluoroethene.
2. Poly(butene).
3. Thermosoftening plastic.
4. (a) Polystyrene packaging, PVC windows, nylon toothbrush bristles.
   (b) Polystyrene is lightweight, absorbs impact, and is a heat insulator; PVC is tough, rigid and waterproof; nylon is strong, flexible, can be drawn into fibres.
5. They are made from crude oil, a non-renewable fossil fuel; they are non-biodegradable so will stay in the environment; they produce poisonous gases when burned.

Ceramics – p69

1. Pottery, tiles, glass, bricks, cement, hair straighteners.
2. Hard-wearing, keep their shape.
3. Heat-resistant and non-stick properties.
4. Thermochromatic plastics – child-friendly thermometers, test strips on batteries, food packaging. Photochromatic plastics – clothing that changes colour or image in sunlight.

Fertilisers – p71

1. Nitrogen, phosphorus, potassium.
2. Animal manure or plant compost.
3. 8%
4. 29% nitrogen, 0% phosphorus
5. KNO_3
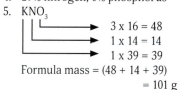
   3 x 16 = 48
   1 x 14 = 14
   1 x 39 = 39
   Formula mass = (48 + 14 + 39)
                = 101 g
   (Total mass of potassium/formula mass of potassium nitrate) x 100
   = (39/101) x 100
   = 38.7%
6. (a) Neutralisation.
   (b) potassium hydroxide + nitric acid → potassium nitrate + water

Nuclear Chemistry – p73

1. (i)
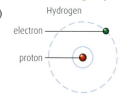
Hydrogen
electron
proton

(ii)
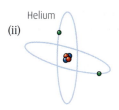
Helium

2.

Background radiation	
Natural sources	**Artificial sources**
Cosmic radiation, including from the Sun	Radioactive waste from nuclear power stations
Food and drink	Nuclear weapons testing and nuclear accidents
Radon gas	Coal-fired and nuclear power stations
Living things	X-rays in medicine

Chemical analysis – p75

1. (a) Pale blue, yellow and red.
   (b) Pencil is insoluble so will not travel up the paper.
2. Potassium.
3. (a) Add a sample of the grit to water; this will dissolve the soluble sodium chloride. Filter the grit and water to remove the sand. Evaporate off the water after filtering, leaving behind solid sodium chloride.
   (b) (i) Dissolve the sodium chloride in water and test with pH paper/universal indicator/pH meter.
       (ii) Use a precipitation reaction and a flame test to prove the presence of sodium and chloride ions.

Chemistry in society: Key area questions

Metals and alloys:
1. (a) A mixture of two or more elements, at least one of which is a metal.
   (b) More resistant to corrosion than iron, gives an attractive shiny finish.
2. (a) Rusting
   (b) Prevents water and oxygen reacting with the iron.
   (c) Zinc
   (d) Painting the metal/covering in oil or grease/covering in plastic/electroplating.

Materials:
1. Thermosoftening means the plastic can be reshaped when heat is applied.
2. (a) Propene
   (b) Lightweight, good heat insulator, unreactive
   (c) Produces poisonous gases

3. (a) A long-chained molecule made up of repeating units called monomers.
    (b) Polymerisation
    (c) They can be made from renewable sources rather than crude oil and will biodegrade so are less harmful to the environment.

Fertilisers:
1. (a) Nitrogen, phosphorus, potassium
    (b) (i) Animal manure or plant compost; (ii) synthetic means artificial.
    (c) It is soluble, a good source of nitrogen.
    (d) 38%

Nuclear chemistry:
1. (a) Cosmic rays/radon gas/food and drink/living plants
    (b) X-rays in medicine/nuclear power stations/smoke detectors/nuclear weapons
    (c) Geiger counter

Chemical analysis:
1. Dissolve a sample of the soil to be tested in water and test the pH using either pH paper or a pH meter. The vegetables whose pH matches that of the soil will grow best.
2. (a) Dark blue sweet, lime green sweet and the dark green sweet.
    (b) Compare the chromatograms of the sweet samples with that of Blueberry blue dye, those with a spot at the same height on the chromatogram as Blueberry blue contain that dye.

Everyday laboratory skills 1 – p79

1. (a) 4.88 g
    (b) The balance could have been TARED.
2. (a) A technique that separates mixtures with different boiling points by heating the mixture until boiling.
    (b) The change of state from a liquid to a gas.
    (c) A piece of equipment that allows a gas to condense back to a liquid.
    (d) The liquid that has been separated from the mixture by evaporation, condensing and has then been collected.
3. (a) As it does not contain flammable compounds.

Everyday laboratory skills 2 – p81

1. (a) Collected in a test tube

gas more dense (heavier) than air

    (b) Collected by displacing water from a test tube

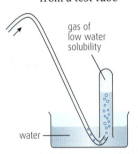

gas of low water solubility

water

    (c) Collected in a test tube

gas more dense (heavier) than air

    (d) Collected in an upturned test tube

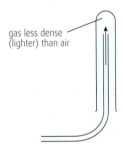

gas less dense (lighter) than air

2. The test tubes of gas could be tested with a burning splint. If the burning splint makes a popping noise when tested with a gas then the gas is hydrogen. The test tubes of gas could then be tested using a glowing splint. If the glowing splint relights when tested with the gas then the gas is oxygen. Finally the gases could be tested with lime water. When lime water is added to the test tube of gas and shaken it will turn cloudy/milky if carbon dioxide is present.

Making a salt – p83

1. (a) Copper sulfate
    (b) Potassium nitrate
    (c) Zinc chloride
2. (a) The process used to separate a solid from a liquid using a funnel and filter paper.
    (b) The liquid that collects in the conical flask after filtration.
    (c) The solid that is collected in the filter paper.
3. By evaporation.

Tables and bar graphs – p85

1. Box round the table, grid lines and units in the table headings.
2. Iron = 0.47 V
    Lead = 1.20 V
    Copper = 1.70 V
    Magnesium = 0.77
3.

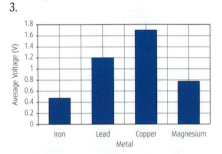

Drawing a line graph – p87

1. (a) time
    (b) temperature
2. A non-linear scale has been used on the y-axis, in other words, the scale does not go up by even amounts. The correct graph is a curve.

Experimental design – p89

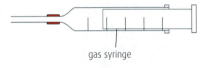

gas syringe

or

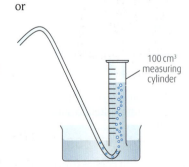

100 cm³ measuring cylinder